An Assessment of the U.S. and Chinese Industrial Bases in Quantum Technology

EDWARD PARKER, DANIEL GONZALES, AJAY K. KOCHHAR,
SYDNEY LITTERER, KATHRYN O'CONNOR, JON SCHMID,
KELLER SCHOLL, RICHARD SILBERGLITT, JOAN CHANG,
CHRISTOPHER A. EUSEBI, SCOTT W. HAROLD

Sponsored by Office of the Undersecretary of Defense for Research
and Engineering
Prepared for Office of the Secretary of Defense
Approved for public release; distribution unlimited

NATIONAL SECURITY RESEARCH DIVISION

For more information on this publication, visit **www.rand.org/t/RRA869-1**.

About RAND

The RAND Corporation is a research organization that develops solutions to public policy challenges to help make communities throughout the world safer and more secure, healthier and more prosperous. RAND is nonprofit, nonpartisan, and committed to the public interest. To learn more about RAND, visit www.rand.org.

Research Integrity

Our mission to help improve policy and decisionmaking through research and analysis is enabled through our core values of quality and objectivity and our unwavering commitment to the highest level of integrity and ethical behavior. To help ensure our research and analysis are rigorous, objective, and nonpartisan, we subject our research publications to a robust and exacting quality-assurance process; avoid both the appearance and reality of financial and other conflicts of interest through staff training, project screening, and a policy of mandatory disclosure; and pursue transparency in our research engagements through our commitment to the open publication of our research findings and recommendations, disclosure of the source of funding of published research, and policies to ensure intellectual independence. For more information, visit www.rand.org/about/principles.

RAND's publications do not necessarily reflect the opinions of its research clients and sponsors.

Published by the RAND Corporation, Santa Monica, Calif.
© 2022 RAND Corporation
RAND® is a registered trademark.

Library of Congress Cataloging-in-Publication Data is available for this publication.

ISBN: 978-1-9774-0826-6

Cover: Bartek Wróblewski / Adobe Stock, Data—liulolo / Getty Images.

About This Report

The authors of this report develop a flexible and repeatable set of metrics for assessing a nation's industrial base in quantum technology—specifically, the nation's scientific research, government support, private industry activity, and technical achievement. The authors apply these metrics to the United States in order to give a holistic assessment of the current state of the U.S. industrial base, and then they apply most of the metrics to the People's Republic of China as a comparative case study. The report concludes with recommendations for policy-makers for preserving the strength of the U.S. industrial base in quantum technology.

The research reported here was completed in October 2021 and underwent security review with the sponsor and the Defense Office of Prepublication and Security Review before public release.

RAND National Security Research Division

This research was sponsored by the Director for Technology and Manufacturing Industrial Base in the Office of the Undersecretary of Defense for Research and Engineering and conducted within the Acquisition and Technology Policy Center of the RAND National Security Research Division (NSRD), which operates the National Defense Research Institute (NDRI), a federally funded research and development center sponsored by the Office of the Secretary of Defense, the Joint Staff, the Unified Combatant Commands, the Navy, the Marine Corps, the defense agencies, and the defense intelligence enterprise.

For more information on the RAND Acquisition and Technology Policy Center, see www.rand.org/nsrd/atp or contact the director (contact information is provided on the webpage).

Acknowledgments

We thank our project sponsor, Robert Gold, and the other key project stakeholders, Paul Lopata, former Principal Director for Quantum Science; and Bethany Harrington, Director for Technology and Industrial Base Assessments. We also thank Ray Mijares, Clare Mernagh, Eun Oh of OUSD (R&E), and Cheryl Samuel of the U.S. Air Force's Office of Commercial and Economic Analysis. All provided invaluable guidance during this project. Within RAND, we thank Howard Shatz, Nathan Beauchamp-Mustafaga, Benjamin Sakarin, and Jim R. Mignano for research assistance; Marjory S. Blumenthal and Caolionn O'Connell for providing very useful feedback as quality assurance reviewers; and Rosa Maria Torres for helping format this report. Finally, we thank the representatives of the industry organizations listed in Chapter Three for agreeing to speak with us.

Summary

There are a variety of early-stage quantum technologies that could eventually deliver major new capabilities to the Department of Defense and other government agencies. Until recently, quantum information science was almost entirely an academic research enterprise, but the private sector has now begun investing in research and development (R&D) as well. Given the newness of private-sector investment in this technology and the high uncertainty in the eventual applications and their timelines, it is difficult to make a holistic assessment of the industrial base in quantum technology. This report develops a set of metrics for assessing a nation's overall quantum industrial base (QIB), broadly defined, applies them to the United States and to the People's Republic of China, and concludes with recommendations for policymakers.

We begin by discussing three broad strategic goals of the department for the domestic (and allied-nation) QIB: quantum technology leadership, quantum technology availability, and the financial stability of QIB firms. Starting from these strategic goals, we formulate 31 specific metrics for assessing a nation's progress toward achieving them. These metrics are designed to be broadly applicable to the United States, to an allied nation, or to a strategic competitor. We focused our research effort on a detailed assessment of these metrics for the U.S. QIB but also apply them to the Chinese QIB as a comparative case study. Whenever possible, we break down the results of our assessment across the three major application domains for quantum technology: quantum computing, quantum communications, and quantum sensing.

Our metrics are divided into four categories, and we took a mixed-method approach to applying them:

1. Our **research metrics** assess the overall output of academia and other producers of open scientific research—for example, the number and growth of total publications within each application domain, the number of publications with high global scientific impact, and the degree of alignment between research topics and policymakers' priorities. Our primary methodology for assessing these metrics was a big-data analysis of virtually all scientific publications in quantum information science (QIS) and technology worldwide over the past decade.
2. Our **government activity metrics** assess the national government's support for R&D in quantum technology—for example, the total amount of R&D funding, its stability over time, and the number of distinct funding sources. Our primary assessment methodologies for these metrics were (a) a review of government policy documents, academic literature, and Chinese-language news sources; and (b) a large-scale analysis of the funding acknowledgment fields in scientific publications.
3. Our **private industry metrics** give an economic overview of the private quantum technology sector as a whole—for example, the total number of firms, their sizes, ages, and funding levels, and any foreign supply-chain dependencies for critical com-

ponents. Our primary assessment methodologies for these metrics were (a) individually profiling a large sample of over 150 quantum technology firms, (b) a review of English- and Chinese-language news and financial reporting, and (c) conversations with subject-matter experts from nine industry organizations.

4. Our **technical metrics** assess the global technical state of the art and innovation potential for specific key technologies. These cross-cutting metrics combine demonstrations by academia, national laboratories, and private industry, so they do not fit cleanly into the previous categories, and they include both sector-wide assessments and assessments of specific technology demonstrations by individual organizations. Our primary assessment methodologies for these metrics were (a) an analysis of patenting activity and (b) a review of the technical scientific literature and other public announcements.

Our key findings on the U.S. and Chinese QIBs are presented here.

The United States' overall scientific research output in quantum information science is broad, stable, and at or near the global forefront in every application domain. The United States has a very broad base of academic research, with over 1,500 research institutions producing more than 10,000 papers over the past decade (focusing on quantum computing most, then communications, then sensing). Publishing in all three domains has seen steady growth of around 12 percent per year. For the output of publications with high scientific impact (as quantified by academic citations), the United States is first in the world in the domains of quantum computing and sensing, and second (after China) in quantum communications. Its research is highly international, with about half of all publications being international collaborations.

The U.S. government is the primary funder of open QIS research and is on track to spend $710 million on QIS R&D in fiscal year (FY) 2021 across multiple agencies. Spending has grown at a rapid rate of about 20 percent per year in recent years, largely driven by a 2018 law known as the National Quantum Initiative.

U.S. quantum technology deployment is now driven by the private sector. U.S. private industry in QIS is broad and diverse, with at least 182 firms—a mixture of large technology companies and recently founded start-ups—pursuing a wide variety of technology approaches and applications, with no single clear technical leader. Venture capital (VC) is a very important source of financing for the start-ups, with $1.28 billion announced so far—the large majority of which has gone to just three firms. Despite the potentially long timelines until useful applications, U.S. private industry is primarily focused on quantum computing, with almost half of the companies and nearly all of the VC investment going toward that domain. Some U.S. companies are dependent on a small number of European or Japanese suppliers for high-quality optical components, but this dependence only applies to companies pursuing certain technical approaches. We did not identify any critical supply-chain dependencies on strategic competitor nations.

The United States leads in demonstrated technical capability in quantum computing and sensing but not in quantum communications. Until recently, the United States was the clear technical leader in every scientific approach to quantum computing. In late June 2021, a Chinese research group claimed technical performance comparable to the United States' in one of the leading scientific approaches (based on superconducting qubits), but their claims are still undergoing peer review as of August 2021. The United States remains the clear leader in most other approaches to quantum computing and also leads in the deployment of quantum sensing. But its R&D in quantum communications remains primarily academic, and the United States lags in deployment—possibly because the only quantum communications application demonstrated so far, quantum key distribution, does not have any clear utility.

Like the United States, China has high research output in every application domain of quantum technology, with more than 14,000 publications over the last decade from over 2,000 research institutions. This research output is growing at about 16 percent per year—somewhat faster than the United States'. China has produced a higher number of highly cited publications in quantum communications than any other nation and is second (behind the United States) in quantum computing and sensing.

Chinese reports of total government R&D funding for quantum technology are wildly conflicting, with publicly reported estimates ranging from $84 million per year to almost $3 billion per year. We are unable to determine from public sources whether the Chinese government is spending more or less than the U.S. government on quantum R&D funding.

Unlike the United States, Chinese quantum R&D is concentrated in government-funded laboratories, which have demonstrated rapid technical progress. Over the past few years, R&D has become heavily concentrated into a single national laboratory in the city of Hefei, which Chinese-language news reports claim is receiving massive funding. By contrast, private firms appear to be much less important players in the Chinese QIB. Although we identified 13 Chinese private companies attempting to deploy quantum technology, they have reported a total of only $44 million in capital funding—only 3 percent of the U.S. private quantum-industry total. Moreover, they have not announced many technically impressive breakthroughs.

China leads in demonstrated technical capability in quantum communications, having launched the world's only quantum communications satellite and being the only nation to demonstrate certain key enabling steps toward the long-distance networking of quantum systems. Several independent lines of evidence indicate that Chinese R&D is focused much more on quantum communications than U.S. R&D is, including the proportions of academic publication and patenting activity, the distribution of private firms, demonstrated technical capabilities, and statements from national leaders. However, Chinese researchers dedicate a higher proportion of their quantum communications research output than U.S. researchers do toward a specific application—quantum key distribution—that both the Defense Science Board and the National Security Agency have publicly announced is not a deployment priority for the U.S. Department of Defense. If their recent claims are verified, then the Chinese are at rough technical parity with the United States in one of the most mature scientific approaches to quantum computing, based on superconducting qubits. China remains behind the United States in most other approaches to quantum computing but is still ahead of the rest of the world. It lags both the United States and Europe in the useful deployment of quantum sensing. Almost all of its most impressive recent technical demonstrations have come from the Hefei national laboratory, reiterating the central role that this laboratory plays in the Chinese QIB.

The eventual applications of quantum technology and their timelines remain highly uncertain, particularly in quantum computing and communications. Many of these technologies are still being advanced through open scientific research that is highly international. We assess that imposing export controls on quantum computing and communications technology would slow scientific progress, and given the early stage of the technology, export controls cannot yet be applied in a way that is targeted to defense-relevant applications.

We conclude with six recommendations for policymakers:

1. Continue to provide a broad base of government R&D support across quantum technologies, complementing the most active areas of private investment.
2. Monitor, and if possible, help protect, the quantum technology programs of key U.S. quantum technology firms.
3. Monitor the financial health and ownership of quantum start-up companies.
4. Monitor the international flows of key elements of the industrial base, such as critical components and materials, skilled workers, and final quantum technology products.
5. Do not impose export controls on quantum computers or quantum communications systems at this time.
6. Periodically reassess the rapidly changing quantum industrial base.

Contents

Figures and Tables

Figures

Tables

CHAPTER ONE

Introduction

In 1982, theoretical physicist Richard Feynman predicted that it should be possible to develop a new type of computer based on quantum phenomena—what is now termed a quantum computer.[1] This insight marked the beginning of an era of rapid theoretical, and later experimental, progress in harnessing the counterintuitive properties of quantum physics for practical purposes. Over the last several decades, researchers have shown that at least in principle, it should be possible to build quantum computers, quantum sensors, and quantum communications systems that can someday provide capabilities far superior to those of systems based on classical physics (such as digital computers).

Over the past decade, academic groups and (more recently) technology companies have demonstrated prototypes of these systems capable of processing certain forms of information. Although these current-generation systems have limited utility because of engineering challenges such as low stability, if technical progress continues, then they may someday realize Feynman's predictions of the ability to sense, process, and transmit far more information than can be achieved using present-day systems.

At the same time, policymakers have become increasingly concerned about the robustness, integrity, and status of the U.S. industrial base, especially those sectors that develop advanced technologies and own unique intellectual property. Such firms, which include defense contractors, have been subject to economic coercion, espionage, and especially cyberattacks by strategic competitor nations.[2] These attacks have enabled strategic competitor nations to acquire advanced technologies useful for their defense programs and to kick-start the development of commercial companies that have quickly gained market share and can compete effectively against companies from U.S.-allied nations.

Even though quantum technologies are still in the early stages of development, it is important to identify and track these technologies, and the key companies and institutions developing them, to ensure this emerging technology base remains secure and in U.S. hands. The United States is currently the overall world leader in quantum technology, but

---

[1] Richard P. Feynman, "Simulating Physics with Computers," *International Journal of Theoretical Physics*, Vol. 21, 1982.

[2] Daniel Gonzales, Sarah Harting, Mary Kate Adgie, Julia Brackup, Lindsey Polley, and Karlyn D. Stanley, *Unclassified and Secure: A Defense Industrial Base Cyber Protection Program for Unclassified Defense Networks*, Santa Monica, Calif.: RAND Corporation, RR-4227-RC, 2020.

this lead is tentative, and the United States should be ready to respond if circumstances change suddenly.

Sectors of the Quantum Industrial Base

Although the term "industrial base" is often used in the context of defense, it more generally refers to a nation's total domestic production capacity (direct or indirect) for a specific industrial sector. For example, the United States has an industrial base for the semiconductor industry.[3] For a mature industry, the industrial base is typically primarily composed of private companies, although there may also be important interconnections with academic laboratories and nonprofit research and development (R&D) organizations as well.[4] But for an early-stage technology like quantum, the relevant components of the industrial base are broader, as they include a higher proportion of organizations specializing in scientific and technology discovery as well as production. We have identified three partially overlapping key sectors of the quantum industrial base (QIB).

Academia and the National Laboratories

The principal contribution of academia to the QIB is basic research that is either of pure scientific interest or at the early stages of applications, with a goal of open publication in the scientific literature. Until recently, quantum information science (QIS) was an almost entirely academic discipline.[5] (In this report, we use the term "academia" broadly to include organizations, including parts of the national laboratories, whose primary output is open scientific research as opposed to commercial products.) As we will demonstrate below, QIS is still an extremely active area of scientific research, and academic institutions are very important for advancing the state of the art, with most key advances documented in detail in the scientific literature. National scientific research capacity is therefore a major driver of technical progress in this field.

Government

The government plays an important role in promoting nascent technology industries by funding scientific research. As with many other scientific areas, most basic U.S. science research

---

[3] Semiconductor Industry Association, "Strengthening the U.S. Semiconductor Industrial Base," webpage, undated.

[4] This is true of the United States. In some other countries, state-owned enterprises play a much larger role in production.

[5] We discuss the subtle distinction between QIS and quantum technology in the next section.

in QIS is funded by government, with federal QIS R&D funding coming from three different "pillars":[6]

- civilian agencies (primarily the National Science Foundation [NSF] and Department of Energy [DoE], and to a lesser extent the National Institute of Standards and Technology [NIST] and National Aeronautics and Space Administration [NASA])
- defense agencies (e.g., the Service labs, Defense Advanced Research Projects Industry [DARPA], and other organizations within the Office of the Secretary of Defense [OSD])
- the intelligence community (with publicly acknowledged funding from the National Security Industry [NSA] and Intelligence Advanced Research Projects Activity [IARPA]).

Through its funding and direct research, the government can play a key role in both accelerating scientific innovation and shaping the academic research portfolio to ensure that its strategic priorities are being addressed.

Industry

The primary functional contribution of industry to the QIB is commercialization. Once a technology reaches sufficient maturity that a commercial market appears imminent, late-stage development and deployment (i.e., scaling and commercialization) usually shifts from academia to the private sector. At this point, many of the later-stage technical developments are protected via trade secrets or by patenting. In the past few years, this shift has begun to occur with QIS, with many companies (both start-ups and established firms) beginning to work on quantum technology and file patents for their inventions. The final applications of these technologies remain highly uncertain, and few of these companies have announced any sales of final products.[7]

These three sectors of the QIB are tightly intertwined in the United States. Most major companies at the forefront of quantum technology were founded by or have chief technology officers who were previously university professors. We have also found significant levels of copublishing in academic journals between private enterprises and university labs. Most key firms belong to one or more cross-sector consortia with academic, government, and industry membership, such as the Quantum Economic Development Consortium (founded by NIST),

---

[6] The government also directly performs its own laboratory research as well as funding academic institutions. We have chosen to include these efforts, which lie in the intersection of government and academia, in the "academia" bin if they produce public scientific findings. "National Quantum Initiative Supplement to the President's FY 2021 Budget," Subcommittee on Quantum Information Science, Committee on Science of the National Science & Technology Council, January 2021.

[7] There are two main exceptions. First, a specific quantum communications technology known as quantum key distribution (QKD) has been commercially deployed in Europe since 2007 and has since been deployed in China, Japan, and South Korea as well. Second, atomic clocks have been in use since the construction of the first cesium atomic clock in 1955 and greatly improved in precision with the invention of laser cooling in the late 1990s.

the Quantum AI Lab (jointly founded by Google and NASA), or the Alliance for Quantum Technologies (jointly founded by Caltech and AT&T Foundry). Such cross-sector linkages have been shown empirically to be critical to a healthy innovative ecosystem.[8]

Overview of Quantum Technologies

Quantum science is a field of physics that studies the physical behavior of microscopic particles (roughly) at or below the scale of individual atoms. At these tiny length scales, particles exhibit behavior that is profoundly different from our everyday experience. For example, a particle no longer has a definite position in space but can only be described as having certain probabilities of being observed at certain locations—or, more loosely, the particle can simultaneously be in many places at once. For decades, these exotic behaviors were considered to be of primarily academic interest. But beginning with Richard Feynman's early discussions of quantum computing in the mid-1980s mentioned previously, the scientific community has gradually come to realize that with careful engineering, sophisticated devices can leverage these phenomena for practical purposes—and in fact can achieve capabilities far beyond any currently existing technologies.

In this report, we use two closely related terms to describe applications of quantum physics. *Quantum information science* refers to the general study (whether theoretical or experimental, purely scientific or applied) of the ways in which quantum physics can be leveraged to gather, process, store, communicate, or protect information. *Quantum technology* refers to practical application of quantum physics in useful devices or algorithms. Government policymakers are primarily interested in the intersection of these two concepts. The proposed applications at this intersection can be grouped into three broad (and somewhat overlapping) application domains: quantum computing, quantum communications, and quantum sensing.[9] We give a short and (relatively) nontechnical overview of these application domains

---

[8] Henry Etzkowitz and Chunyan Zhou, *The Triple Helix: University-Industry-Government Innovation and Entrepreneurship*, London: Routledge, 2017. Jon Schmid, Sergey A. Kolesnikov, and Jan Youtie, "Plans Versus Experiences in Transitioning Transnational Education into Research and Economic Development: A Case Study," *Science and Public Policy*, Vol. 45, No. 1, 2018.

[9] It is always challenging to draw a sharp distinction between what counts as a "quantum technology" and what does not. Quantum theory first began development in 1905, and it has been leveraged for practical applications going back at least as far as the 1947 invention of the transistor, whose behavior can only be understood using quantum concepts. In this report, we reserve the term "quantum technology" to describe technologies in which quantum particles can be precisely controlled and measured at the *individual* level. It does not include devices that can be understood and engineered entirely using traditional materials science, which considers the overall statistical behavior of huge numbers of quantum particles but not the individual behavior of single particles. (However, the distinction is sometimes blurry, particularly in quantum sensing.) Our much more recent capability to control quantum particles at the individual level is sometimes called the "second quantum revolution." Jonathan P. Dowling and Gerard J. Milburn, "Quantum Technology: The Second Quantum Revolution," *Philosophical Transactions of the Royal Society A*, Vol. 361, 2003.

below, focusing on their current status and possible applications rather than on the underlying science.

Quantum Computing

A *quantum computer* is a physical computer that leverages the unique properties of quantum physics to operate on fundamentally different principles from essentially all existing computers (which are collectively referred to as *classical* as opposed to *quantum* computers). The smallest elementary building block of a classical computer is a *bit*, which is any physical device with two states that represent either a logical 0 or a logical 1. By contrast, the elementary building block of a quantum computer is a *qubit*. A qubit is a (sometimes microscopic) physical quantum system that can similarly be in either of two states that represent a logical 0 or 1—but they can also be in a *superposition* of the 0 and 1 states. Loosely, this means that each qubit can be thought of as simultaneously existing in both the 0 *and* 1 states at the same time.

A single qubit is not particularly useful; the true utility of qubits occurs when many qubits are combined together. When arranged properly, a collection of multiple qubits can be jointly placed into a collective superposition—for example, if we think of three physical qubits as representing a logical string of three bits, then the three qubits can exist in a single superposition that simultaneously represents the eight bitstrings 000, 001, 010, 011, 100, 101, 110, and 111.

When multiple qubits jointly enter into such a superposition, they are referred to as being *entangled* together.[10] Entanglement is the property that unlocks the power of quantum computers: loosely speaking, if a quantum computer's qubits are entangled together, then it can mathematically process all of the corresponding bitstrings simultaneously, whereas a classical computer is forced to process each bitstring one at a time. A quantum computer can therefore be thought of as being capable of organically performing massively parallel computations.[11] Moreover, when multiple qubits are entangled together, the number of states in which

---

Timing (e.g., via atomic clocks) is sometimes broken out separately from other types of quantum sensing. Although this report does not focus on atomic clocks in depth, we have chosen to group timing as a part of quantum sensing, in accordance with the National Quantum Initiative's Program Component Areas.

[10] Strictly speaking, the qubits are only actually entangled together if (roughly speaking) the marginal probabilities of measuring each individual qubit to be in a 0 or 1 state are correlated. In this case, we cannot describe the state of each qubit individually but must instead describe the superposition state of the entire collection of qubits.

[11] The common explanation presented here is highly simplified. Some quantum computing experts dislike this explanation and believe that it is oversimplified to the point of being actively misleading. It omits many key details regarding the probabilistic nature of a quantum state that is encoded in the detailed nature of the superposition, which places severe limits on how much of this notional "parallelism" can actually be exploited in practice. Scott Aaronson, *Quantum Computing Since Democritus*, Cambridge: Cambridge University Press, 2013. We agree that the explanation given above is not very accurate for explaining the technical functioning of a quantum computer, but we think it is appropriate for the goals of this report.

they can simultaneously exist scales exponentially with the number of qubits. Very roughly, each additional qubit *doubles* the memory of the quantum computer, while each additional bit increases the memory of a classical computer by only a fixed amount.

For *some* applications, this effective parallelism allows quantum computers to perform calculations vastly faster than any existing computers. However, quantum computers operate on very different logical principles from regular computers, and completely new quantum algorithms must be developed for different applications. Although many different quantum algorithms have been developed, the most important ones fall into four major categories (listed in roughly increasing order of technical difficulty, although there is some uncertainty there):

1. There are several quantum algorithms for modeling and simulation as applied to biochemistry and materials science.[12] Possible applications include drug discovery, agriculture, and the design of new advanced materials (e.g., improved solar panels or airframes). This is the application that many experts believe to be the mostly likely to be realized first.[13]

2. A quantum algorithm known as *Grover's algorithm* speeds up the calculation of a broad class of challenging search and optimization problems for which the best-known algorithms involve brute-force search.[14] Although Grover's algorithm is very broadly applicable, the speedup that it delivers is relatively modest: it provides some improvement but does not by any means make these computations easy.[15] There are a huge number of potential applications to computationally intensive combinatorial problems such as transportation logistics, weather modeling, and fluid and other scientific simulations.

3. The most famous quantum algorithm, *Shor's algorithm*, allows a quantum computer to rapidly factor large numbers.[16] Almost all public-key encryption systems in use today rely on the assumption that factoring (and related problems) are computation-

---

[12] Markus Reiher et al., "Elucidating Reaction Mechanisms on Quantum Computers," *Proceedings of the National Academy of Sciences*, Vol. 114, No. 29, 2017.

[13] Katerine Bourzac, "Chemistry Is Quantum Computing's Killer App," *Chemical & Engineering News*, Vol. 95, No. 43, October 30, 2017.

[14] Lov K. Grover, "A Fast Quantum Mechanical Algorithm for Database Search," *28th Annual ACM Symposium on the Theory of Computing*, 1996.

[15] Roughly speaking, if a brute-force classical search requires N time steps, then Grover's algorithm can reduce this to $\sqrt{N}$ time steps on a quantum computer. Unlike for materials simulation and Shor's algorithm, this improvement is not exponential or even superpolynomial. So if the combinatorial problem is exponentially difficult, then it remains so even if Grover's algorithm is applied.

[16] Unlike Grover's algorithm, Shor's algorithm provides a huge (almost exponential) speedup over the best-known classical algorithm for factoring. It therefore could change the mathematical task of factoring from intractable to easy. Peter W. Shor, "Polynomial-Time Algorithms for Prime Factorization and Discrete Logarithms on a Quantum Computer," *SIAM Journal on Computing*, Vol. 26, No. 5, 1997.

ally intractable, so a large-scale quantum computer running Shor's algorithm could eventually easily decrypt most of the information transmitted over the internet, which poses clear threats to both the United States' national and economic security.[17]

4. A variety of quantum algorithms have been proposed for machine learning in artificial intelligence (AI), most prominently the *Harrow-Hassidim-Lloyd (HHL) algorithm*.[18] This area is under very active research, but the final applications are not yet as well understood as the previous three. There have been many useful partial discoveries but still no end-to-end quantum algorithm that demonstrates a clear speedup over classical algorithms. So the prospects are still unclear.[19]

There may well be other powerful quantum computing algorithms that have not yet been discovered. The theoretical capabilities of quantum computers are still only partially understood, in part because it is difficult to study a quantum computer purely theoretically without any actual hardware to experiment on. However, it is strongly believed that quantum computers will give a huge speedup over classical computers for only certain types of math problems; for other problems, they will deliver only modest advantages.[20]

Building a quantum computer is very technically challenging because entangled qubits are extremely unstable. Even tiny disturbances cause qubits to *decohere*, or drop out of the quantum superposition that is the key to their operation. The qubits therefore need to be kept extremely well isolated from all environmental disturbances, which often requires them to be kept in extremely strong vacuums or cooled down to a fraction of a degree above absolute zero. There are a variety of different physical architectures for qubits under development:[21]

1. superconducting-transmon qubits, which are tiny chips of superconducting metals cooled to about one-thousandth of a degree above absolute zero

2. trapped-ion qubits, which are individual atoms suspended in place by lasers in a strong vacuum; a close variant are neutral cold-atom qubits[22]

---

[17] Michael J. D. Vermeer and Evan D. Peet, *Securing Communications in the Quantum Computing Age: Managing the Risks to Encryption*, Santa Monica, Calif.: RAND Corporation, RR-3102-RC, 2020.

[18] Aram W. Harrow, Avinatan Hassidim, and Seth Lloyd, "Quantum Algorithm for Linear Systems of Equations," *Physical Review Letters*, Vol. 103, No. 15, 2009.

[19] Scott Aaronson, "Read the Fine Print," *Nature Physics*, Vol. 11, 2015; Ewin Tang, "Quantum Principal Component Analysis Only Achieves an Exponential Speedup Because of Its State Preparation Assumptions," *Physical Review Letters*, Vol. 127, 2021, p. 060503.

[20] For example, contrary to frequent claims, quantum computers are *not* believed to be able to deliver exponential speedups—even in principle—for a challenging class of math problems known as NP-complete problems (such as the traveling salesperson problem). Quantum computers are believed to be able to deliver only modest speedups for these problems; see Aaronson, 2013.

[21] T. Ladd et al., "Quantum Computers," *Nature*, Vol. 464, 2010.

[22] The difference is that in the ion case, the atoms have had an electron removed, so they have a net positive electric charge, which makes them easier to hold in place but harder to keep isolated.

3. photonic (or optical) qubits, which are individual particles of light known as *photons*
4. quantum-dot (or solid-state or spin) qubits, which are small semiconducting crystals
5. topological qubits, which are made of exotic *topological* states of matter that have not yet been created in the lab; this type of qubit is still only theoretical.

It is not yet clear which of these qubit architectures will work best for practical quantum computers.

In 2019, Google announced that it had built a quantum computer named Sycamore with 53 qubits, which had performed a calculation that was too difficult for even the world's fastest existing classical supercomputer.[23] This milestone is referred to as *quantum supremacy*. The specific math problem that the Sycamore computer solved (known as *random quantum circuit sampling*) is not believed to have any practical applications; this demonstration was merely a proof of principle that quantum computers could exceed the world's best supercomputers for certain problems. As of July 2021, Sycamore is the only quantum computer that has achieved quantum supremacy according to broad academic consensus (although as we discuss later, new claims to quantum supremacy have recently arisen).[24]

All quantum computers that have been demonstrated today have around 64 qubits or fewer, which is not enough for a useful calculation. Moreover, their qubits remain highly unstable and decohere within a fraction of a second, which effectively terminates the calculation. This decoherence greatly limits the utility of existing prototypes. These rudimentary and unstable computers are known as *noisy intermediate-scale quantum* (NISQ) computers. There are currently no confirmed useful algorithms that can run on NISQ computers, although this is an area of very active investigation.

There exists a known theoretical solution to the problem of decoherence, known as *quantum error correction* (QEC). QEC is a technique of networking qubits together and controlling them in such a way that environmental noise can be detected and corrected before it causes the qubits to decohere. QEC is still in the very early stages of development; only very recently has it been demonstrated in a limited capacity on a small number of qubits.[25] QEC is critical to unlocking the full potential of quantum computers; all of the major applications discussed above are so complex that they will require extensive QEC for useful applications.[26]

---

[23] F. Arute et al., "Quantum Supremacy Using a Programmable Superconducting Processor," *Nature*, Vol. 574, 2019.

[24] There is some dispute in the expert community as to just how *much* faster Sycamore ran than the fastest classical supercomputer. However, given the lack of any practical application of the math problem that it solved, the exact magnitude of the speedup is not terribly important as a practical matter.

[25] Google Quantum AI, "Exponential Suppression of Bit or Phase Errors with Cyclic Error Correction," *Nature*, Vol. 595, 2021; Laird Egan et al., "Fault-Tolerant Control of an Error-Corrected Qubit," *Nature*, Vol. 598, 2021.

[26] There have, however, been proposals for achieving more limited applications in chemistry simulation or combinatorial optimization on NISQ computers. Abhinav Kandala et al., "Hardware-Efficient Variational Quantum Eigensolver for Small Molecules and Quantum Magnets," *Nature*, Vol. 549, 2017.

Unfortunately, QEC requires a huge qubit overhead; it is estimated that implementing QEC could increase the number of qubits required for an application by a factor of at least 1,000. It therefore will not be possible to fully implement QEC until thousands of qubits have been successfully networked together.

The timeline for useful applications of quantum computers is therefore still extremely uncertain. However, an expert panel from the National Academy of Sciences concluded in 2019 that a quantum computer capable of threatening encryption was still at least a decade away.[27]

The quantum computers discussed above all belong to the most theoretically powerful class, known as *universal quantum computers*. But there also exist other, more limited forms of quantum computing.

One alternative approach is known as *quantum annealing* and is used for numerical combinatorial optimization of complex functions. This was the first type of quantum computing to be deployed commercially (by a Canadian company known as D-Wave in 2011). The main advantage of quantum annealing is that it may be more robust to some level of noise and hardware errors in the system, but the exact computational power of quantum annealing is unclear, both in principle and in practice. The general consensus among experts is that the most advanced quantum annealer (made by D-Wave) has not yet demonstrated any clear and scalable computational advantage over existing classical computers.[28]

A second alternative is known as *boson sampling*.[29] Unlike standard (or *universal*) quantum computing, boson sampling does not use qubits but instead microscopic particles known as *bosons* (typically photons). Large numbers of bosons interact in complex but mathematically well-specified ways that are believed to be intractable to simulate on a classical computer, thereby providing a form of computation. Boson sampling cannot perform all of the types of mathematical calculations that a universal quantum computer (or even a classical computer) can. There have been a few proposals for practically useful calculations using boson sampling, but (unlike with universal quantum computing) there is no consensus within the scientific community that boson sampling will ever deliver any practical advantages over a classical computer, even in principle.

A third alternative is a type of *quantum simulation* in which an array of qubits (or similar systems) directly mimics the physical arrangement of atoms in a solid.[30] That is, rather than using the qubits as abstract logic units, each qubit directly represents a single atom, and

---

[27] National Academies of Sciences, Engineering, and Medicine, *Quantum Computing: Progress and Prospects*, Washington, D.C.: National Academies Press, 2019.

[28] Vasil S. Denchev et al., "What Is the Computational Value of Finite-Range Tunneling?" *Physical Review X*, Vol. 6, No. 3, 2016.

[29] Scott Aaronson and Alex Arkhipov, "The Computational Complexity of Linear Optics," *Theory of Computing*, Vol. 9, No. 4, February 2013.

[30] Sepehr Ebadi et al., "Quantum Phases of Matter on a 256-Atom Programmable Quantum Simulator," *Nature*, Vol. 595, No. 7866, July 2021.

the interactions between nearby qubits are made to approximate the physical interactions between nearby atoms in a real material. Although less general than universal quantum computing, this approach has many real-world scientific applications.

Quantum Communications

Quantum communications refers to any technology that physically transmits a quantum state (encoding useful information) over a significant distance. The medium of transmission is almost always a stream of photons (particles of light). In principle, these can be transmitted through all of the same channels as conventional electromagnetic signals: fiber-optic cable, open air, satellite, underwater, etc. In practice, photons correspond to extremely weak and highly directional signals, making transmission technically challenging.

There are two main "generations" of quantum communications technology, with different applications. "First-generation" quantum communications does not use the phenomenon of quantum entanglement discussed above; the photons are sent out individually. The primary application of this technology is a transmission technique known as quantum key distribution (QKD), a type of *quantum cryptography*, which in principle improves the security of the transmission against interception. In QKD, a stream of photons is engineered in such a way that due to the counterintuitive behavior of quantum particles, any eavesdropper that intercepts the transmission will (in principle) *inevitably* leave a signature of their interception on the signal itself. If the recipient detects such an interference, then they can discard the transmitted information and inform the sender to start the transmission over.[31] QKD does not directly reduce the risk of transmission detection or degradation but only of interception. It has been commercially available in Europe since 2007; its deployment is growing rapidly in Europe, China, South Korea, and Japan, but there has been very little deployment in the United States.

QKD provides one possible solution to the threat that quantum computers pose to encryption, as it is not vulnerable to Shor's algorithm or any other quantum attacks. But in practice, it is complex, expensive, and has repeatedly demonstrated implementation errors that lead to subtle security vulnerabilities.[32] As such, the U.S. Department of Defense's (DoD's) Defense Science Board has publicly concluded that "QKD has not been implemented with sufficient capability or security to be deployed for DoD mission use."[33] Likewise, the U.S. NSA has pub-

---

[31] The transmitted data is not sensitive information itself, but simply a string of random bits that is later used as an encryption key to encrypt the sensitive information using a technique known as *symmetric-key encryption*.

[32] Y. Y. Fei et al., "Quantum Man-in-the-Middle Attack on the Calibration Process of Quantum Key Distribution," *International Journal of Scientific Reports*, Vol. 8, No. 4283, 2018; Xiao-Ling Pang et al., "Hacking Quantum Key Distribution via Injection Locking," *Physics Review Applied*, Vol. 13, 2020.

[33] Defense Science Board, "Applications of Quantum Technologies—Executive Summary," October 2019.

licly announced that "NSA does not support the usage of QKD or [quantum cryptography] to protect communications in National Security Systems."[34]

The "second generation" of quantum communications technologies is still in its infancy. It involves *entanglement distribution*—that is, taking two qubits (typically photons) that are entangled together and physically separating them by long distances while maintaining their quantum entanglement. The ability to control far-separated entangled pairs enables a large number of sophisticated quantum processing techniques, such as a process called *quantum teleportation* in which the full quantum state of a qubit is transferred over long distances without needing to physically move the particle.

The main application of entanglement-based quantum communications technology is believed to be *quantum networks* that link together quantum devices such as computers or sensors in a way that allows them to collectively process information in parallel. As a nascent technology, it is not yet clear which precise applications will prove feasible, but the DoD has identified distributed quantum computing, quantum sensor arrays, and clock synchronization as areas of interest.[35] A hypothetical future large-scale network of devices (whether classical or quantum) connected by quantum communications links is sometimes referred to as the *quantum internet*.[36]

Quantum Sensing and Timing

Quantum sensing refers to the use of quantum physics to improve the capabilities of a wide variety of types of sensors—for example, clocks, accelerometers, gyroscopes, gravimeters, antennas and other electromagnetic radiation detectors, and sensors for steady electric or magnetic fields. Unlike quantum computing and communications, quantum sensing generally does not involve conceptually new capabilities, but in some cases, the quantitative improvement is large enough that it may enable new capabilities.

Quantum sensing is arguably the most diverse quantum technology application domain— in terms of potential classes of useful devices, the nature of the quantum improvement, and the relevant technical approaches (ultracold atom gases, room-temperature vapor cells,

---

[34] Instead, the NSA endorses the adoption of a different system for securing sensitive encrypted information against future quantum computers: new cryptographic algorithms known as *post-quantum cryptography* (PQC) that can be implemented on classical computers and are believed (although not mathematically proven) to be secure against attacks from quantum computers. See Vermeer and Peet, 2020, for a discussion of policy challenges surrounding the transition to PQC algorithms. National Security Agency, "Quantum Key Distribution (QKD) and Quantum Cryptography (QC)," undated.

[35] Entanglement-based quantum communications also enable a more sophisticated form of QKD known as *measurement-device-independent* QKD, which closes many of the security loopholes inherent to first-generation QKD. Defense Science Board, 2019.

[36] Natalie Wolchover, "To Invent a Quantum Internet," *Quanta Magazine*, September 25, 2019.

Rydberg atoms,[37] etc.). In some cases, the advantage that quantum sensors deliver over classical ones involves higher sensitivity; in others, it involves lower size, weight, power, and cost requirements or longer stability. As such, it is difficult to summarize the current state or applications of quantum sensors.

However, DoD has publicly identified certain areas of potential interest: quantum accelerometers and gyroscopes for improved inertial guidance, gravimeters for gravity-aided navigation and underground tunnel detection, and miniaturized antennas.[38] Several organizations have announced that they are developing prototype quantum magnetometers for navigation using Earth's magnetic field as a supplement to GPS. Quantum inertial, gravitational, and magnetic sensors together offer several potential venues for improving positioning, navigation, and timing (PNT) capabilities.

Quantum sensors also have applications for intelligence, surveillance, and reconnaissance (ISR), such as the use of Rydberg atoms for sensitive antennas.[39] A less conventional sensing modality that has received recent study is known as *quantum illumination*, which involves generating a pair of entangled phonons, reflecting one photon off a target, and then recombining it with its partner and jointly measuring them. This technique has only been demonstrated at tabletop scale but has been shown to significantly improve the detector's signal-to-noise ratio in noisy environments.[40] Biomedical imaging has been proposed as a potential application of short-distance quantum illumination. A less mature variation of quantum illumination known as *quantum radar* proposes to use radio-frequency entangled photons for long-distance ISR.[41] However, the Defense Science Board has concluded that "quantum radar will not provide upgraded capability to DoD."[42]

Summary

Figure 1.1 summarizes the DoD's assessment of the current military readiness and impact of several of the quantum technologies discussed above.

---

[37] A Rydberg atom is a nearly ionized atom with one electron excited to a high energy level. Because this electron is only loosely bound to the atomic nucleus, it is very sensitive to the presence of external electromagnetic fields.

[38] Defense Science Board, 2019.

[39] Defense Science Board, 2019.

[40] Zheshen Zhang et al., "Entanglement-Enhanced Sensing in a Lossy and Noisy Environment," *Physical Review Letters*, Vol. 114, 2015.

[41] Shabir Barzanjeh, "Microwave Quantum Illumination," *Physical Review Letters*, Vol. 114, 2015.

[42] Defense Science Board, 2019.

FIGURE 1.1

Summary of Military Readiness and Impact of Various Quantum Technologies

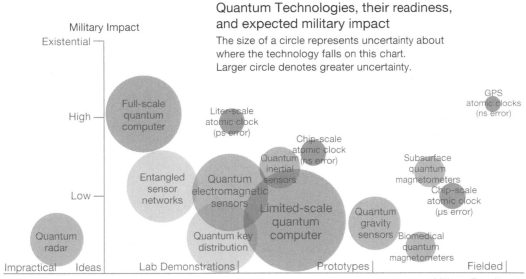

SOURCE: Provided to RAND by the Office of the Undersecretary of Defense for Research and Engineering.
NOTE: This chart updates a previous version published in the Fiscal Year 2020 Industrial Capabilities Report to Congress, 2021.

Objectives of This Assessment

This assessment had four objectives: first, to develop a generally applicable set of measures and metrics for holistically assessing a nation's R&D and industrial capacity in quantum technology; second, to apply these metrics to the United States in order to demonstrate their utility; third, to apply a subset of them to the People's Republic of China (another nation at the forefront of quantum technology) as a comparative case study; and fourth, to use the results of our assessment to identify policy recommendations for strengthening the U.S. quantum industrial base. If these metrics prove useful, then we hope that both the United States and allied nations will continue to reapply them in the future to assess the state of their own industrial bases as they continue to mature.[43]

---

[43] For a somewhat similar assessment that focuses only on quantum computing, see Philipp Gerbert and Frank Ruess, *The Next Decade in Quantum Computing—and How to Play*, Boston, Boston Consulting Group, November 2018.

CHAPTER TWO

Strategic Goals and Metrics for Quantum Industrial Base Assessment

Strategic Goals for the Quantum Industrial Base

One of the U.S. DoD's key missions is to "ensure robust, secure, resilient, and innovative industrial capabilities" of the defense industrial base (DIB).[1] Furthermore,

> the national security of the United States requires the technological and intellectual capabilities of domestic and foreign companies, academia, and dual-use technology providers collaborating at the forefront of future generation technologies.[2]

DoD monitors and supports DIB firms so they can provide the technologies needed by DoD and to ensure DoD weapons and other systems are superior to those of strategic competitors. Achieving technological superiority and maintaining this advantage underpin U.S. military strategy and also can provide significant benefits to the U.S. economy. Quantum technology may be a key emerging technology for DoD. This suggests DoD should have the following strategic goals for the QIB:

1. quantum technology leadership
2. quantum technology availability
3. financial stability of QIB firms.

Given the long-term horizons and high uncertainty for the development of quantum technology, three subgoals are important for ensuring technology leadership:

- The nation's academia and industry should lead internationally in developing new quantum technologies and systems.

---

[1] Office of the Undersecretary of Defense for Acquisition and Sustainment (OSD A&S) Industrial Policy, *Fiscal Year 2019 Industrial Capabilities Report to Congress*, June 23, 2020.

[2] *Fiscal Year 2019, Industrial Capabilities Report to Congress*, 2020.

- The QIB should demonstrate sustained progress in developing quantum technologies.
- Because quantum technology development is in its early stages, the QIB should support a diversified portfolio of early-stage quantum technologies. Some will have uncertain payoffs, and some may not be successful, but a few will, and these may be the ones that one day provide the warfighting or intelligence capabilities needed by DoD.

Quantum technologies should also be available to defense contractors, the military, and more broadly to the federal government. Three subgoals are important for ensuring technology availability:

- Domestic or allied-nation organizations can deploy integrated quantum systems at useful scale.
- The QIB is not dependent on foreign technologies from strategic competitor nations.
- The QIB provides technologies that meet strategic government priorities.

The third QIB strategic goal focuses on the financial health and stability of the QIB—both firms that produce key quantum technologies and academic institutions that may be developing emerging quantum technologies. The QIB should be stable against the failure of key firms or their acquisition by foreign companies domiciled in strategic competitor nations. Emerging technologies developed by the domestic QIB should be protected, and transfer of such technologies to strategic competitors should be minimized whenever possible.

We used these proposed strategic goals to guide our research and the development of metrics to assess the status and health of the QIB.

Metrics Used in This Report

We developed a set of 31 metrics that attempts to give insight into the current state of the QIB and its likelihood of meeting the strategic goals discussed earlier. (Although the strategic goals were chosen from the perspective of U.S. DoD policymakers, the metrics are designed to be broadly applicable to any nation, including both allied nations and strategic competitors.) We tried to be as comprehensive as possible while keeping the metrics feasible to assess in practice.[3] Some of these metrics are very precisely defined, while others require subject-matter expert (SME) judgment. For most of the numerical metrics, there is a fairly clear correspondence between the value and its alignment with the strategic goals (i.e., either a higher number or a lower number reflects a stronger industrial base), but for some metrics, both higher and lower values have their pros and cons, and the desired direction of movement may depend on policymakers' judgment.

---

[3] Scott Savitz, Miriam Matthews, and Sarah Weilant, *Assessing Impact to Inform Decisions: A Toolkit on Measures for Policymakers*, Santa Monica, Calif.: RAND Corporation, TL-263-OSD, 2017.

We divided our metrics into four top-level categories, denoted by Roman numerals. The first three categories roughly correspond to the three sectors of the QIB discussed earlier (academia, government, and industry).

1. Scientific research metrics: These are gathered from the nation's output of scientific research openly published in scientific journals. They primarily capture the output of the academia and the national laboratories, although they also include some scientific research performed by private companies.
2. Government activity metrics: These capture the nature of government funding support for quantum R&D.
3. Private industry metrics: These give a broad sector-wide characterization of the nation's overall private industry in quantum technology, such as the distribution of firms across various economic dimensions.
4. Technical metrics: These give a more granular and in-depth assessment of the nation's technical state of the art, across both academic research and private industry deployment.

Within each of these four categories, we determined several key high-level aspects of the quantum ecosystem (denoted by capital letters) that reflect the alignment of the nation's QIB with the strategic goals discussed earlier. Finally, we developed specific (mostly quantitative) metrics (denoted by numbers) to quantify the state of each aspect of the ecosystem. The complete set of metrics is summarized in Table 2.1 and discussed in more detail in the next subsection.

In consultation with our research sponsors, we chose not to include any explicit metrics to assess the status of the skilled workforce on quantum technology. This decision is not intended to minimize the critical importance of a skilled workforce to the industrial base; rather, it was solely a matter of research scoping.

The rest of this chapter discusses our chosen metrics in detail and briefly summarizes our assessment methodology. In Chapters Three and Four, we apply these metrics to the two most important national players in quantum technology, the United States and China. We separately assess each metric across the three major quantum application domains of computing, communications, and sensing.

I. Research Metrics
A. Overall Research Activity

Metric I.A.1, the total number of academic publications on quantum technology that the nation publishes, reflects the nation's overall degree of research activity in quantum technology. Even though quantity does not necessarily correspond to quality or impact, the total activity gives a rough indication of how much the nation is focusing on quantum technology and its impact on the advancement of the science, as well as its relative prioritization of the application domains.

TABLE 2.1

QIB Metrics Used in This Report

Aspect of Ecosystem Being Measured	Metrics
I. Research metrics	
A. Overall research activity	1. Total publication count
B. Growth in research activity	1. Annual growth in publication count
C. Institutional concentration of research activity	1. Number of publishing institutions 2. Herfindahl–Hirschman Index (HHI) for publication counts
D. Global scientific impact	1. Percentage of world's highly cited publications 2. Number of institutions producing highly cited research
E. Topical alignment with government priorities	1. Percentage of publications about topics of high or low strategic priority to policymakers
F. Degree of domestic and international research collaboration	1. Average number of collaborating domestic institutions 2. Percentage of publications coauthored with other nations
G. Risk of technology leakage	1. Percentage of publications funded by or coauthored with strategic competitor nations 2. Number of domestic coauthors with strategic competitor military institutions
II. Government activity metrics	
A. Overall government R&D investment	1. Total R&D funding
B. Growth in investment	1. Annual growth in funding
C. Stability of investment	1. Number and length of multiyear funding commitments
D. Breadth of investment sources	1. Number of significant funding sources 2. HHI for funding agencies by resulting publication counts
III. Private industry metrics	
A. Number and distribution of QIB firms	1. Total number of firms 2. Distribution of firms by size 3. Distribution of firms by age 4. Distribution of firms by VC funding or revenue 5. Distribution of firms across the production chain
B. Degree of firm specialization to quantum technology	1. Proportion of quantum-involved firms that are primarily dedicated to quantum technology
C. Foreign supply-chain dependencies	1. Foreign dependencies for critical materials, components, or services 2. Foreign dependencies on strategic competitors
IV. Technical metrics	
A. Innovation potential	1. Number of patent applications filed, by application domain 2. Number of unique patent assignees 3. Growth rate of patent applications
B. Technical achievement	1. [Application-specific technical metrics]
C. Breadth of technical approaches under pursuit	1. Number of distinct technologies in development or production 2. Application subdomains in which nation is a world leader 3. Technologies with a quantitative road map to deployment

B. Growth in Research

Metric I.B.1, the annual relative growth in publication count, captures the growth in research activity over time. For a field like quantum that is still largely in the basic research stage, sustained growth over time indicates that the nation is prioritizing quantum technology over the long term.

C. Institutional Research Capacity

Institutional research capacity quantifies the number of distinct research institutions that are actively researching quantum technology. Centralized and decentralized research models each have their advantages and disadvantages. Having a small number of distinct research institutions reduces the risk of research duplication and siloing, but having a large number of institutions encourages a variety of research approaches and priorities, decreases the risk of a single key institution reducing its research activities and setting back the field, and generally correlates with more researchers working in the field.[4] Given the importance of a diversified technology portfolio, we judge that a large number of research institutions (that are all significant players) is a sign of robustness for the QIB.

Metric **I.C.1** captures the total number of unique institutions (academic, laboratory, or industry) that publish research in QIS during a given time period. Tracked over time, this metric can indicate whether the nation's overall research effort in quantum is growing or shrinking.

Metric **I.C.2** captures the institutional *concentration* of research: whether a small number of institutions are generating the large majority of the research output, or whether the output is fairly uniformly distributed across many institutions. We quantify this using a metric known as the Herfindahl-Hirschman Index (HHI), which is commonly used in the context of antitrust regulation to measure market concentration. As discussed in the Appendix, we have adapted this metric to the institutional publication counts to measure the concentration of the nation's overall quantum research. The HHI is a number between 0 and 1; an HHI close to 1 means that almost all of the research output comes from a single institution, while an HHI close to 0 means that a large number of institutions are all producing a comparable share of the research. Given the benefits of a stable and diversified research portfolio, we believe that (all else equal) a low HHI is a positive sign.

D. Global Scientific Impact

If a nation is a global leader in developing new quantum technologies, then its research activity will strongly impact the rest of the world's R&D as well. The global impact of its research is a proxy for quality, which is just as important as quantity. Though not perfect, the most widely accepted metric of research impact within the academic community is the number of

---

[4] Thomas Bryan Smith, Raffaele Vacca, Till Krenz, and Christopher McCarty, "Great Minds Think Alike, or Do They Often Differ? Research Topic Overlap and the Formation of Scientific Teams," *Journal of Informetrics*, Vol. 15, No. 1, 2021.

citations it receives.[5] **Metric I.D.1** quantifies the nation's scientific impact by counting the number of high-impact academic publications that it generates.[6] **Metric I.D.2**, the number of distinct research units that produce highly cited research, counts the number of major global research institutional "players" in the nation as another measure of global impact.[7]

E. Topical Alignment with Government Priorities

Government policymakers may have strategic priorities for quantum technology (particularly in the national security realm) that deviate from the near-term commercial demand, so even a strong overall national R&D position may not ensure that all of the key quantum technologies are being developed. **Metric I.E.1** quantifies the percentage of research that focuses on technologies that policymakers have identified as being of particularly high or low interest.[8] These specific technologies will change over time as quantum technology matures, but the results in Chapters Three and Four discuss examples of technology that the DoD has publicly identified as lower priority as of 2021.

F. Degree of Domestic and International Collaboration

Scientific collaboration among research institutions has been demonstrated to promote technological development, prevent siloing, and reflect a healthy research base.[9] We have separate metrics to quantify the degree of domestic and international collaboration.

The connectedness of the domestic collaboration network is quantified by **Metric I.F.1**, the average number of collaborating domestic institutions per domestic institution.[10] The higher the number, the higher the level of domestic collaboration.

We quantified the international research collaboration by a simpler **Metric I.F.2**, which is the percentage of domestic publications that have at least one international collaborator. Quantum technology is a highly international topic, with major advances occurring in many

---

[5] Dag W. Aksnes, Liv Langfeldt, and Paul Wouters, "Citations, Citation Indicators, and Research Quality: An Overview of Basic Concepts and Theories," *SAGE Open*, January 2019.

[6] Specifically, the number of publications in the top decile of citations received during a given year.

[7] We define a research unit as a group of geographically colocated researchers that share an organizational affiliation. For a description of why we prefer this unit to affiliation, see the Appendix.

[8] Our project stakeholders conveyed to us that they wanted our set of metrics to be repeatable and broadly applicable by a variety of users. As such, we have for the most part deliberately remained general in our discussions of policymakers, rather than indicating specific institutional stakeholders.

[9] Stefan Wuchty, Benjamin F. Jones, and Brian Uzzi, "The Increasing Dominance of Teams in Production of Knowledge," *Science*, Vol. 316, No. 5827, 2007.

[10] We chose to weight the average by the number of shared publications in order to capture the strength of collaboration; see Appendix A for more details. For countries with a large number of domestic research institutions, such as the United States and China, we believe the average number of domestic collaborators is the best metric for domestic collaboration. But for countries with smaller domestic networks, this number can be limited by the low number of institutions available for collaboration. For these countries, a more useful metric for comparison might be the percentage of possible pairs of institutions that have collaborated.

countries, so we believe that a nation whose research is isolated from the rest of the world is unlikely to be at the global scientific forefront.[11]

G. Risk of Technology Leakage

The international exchange of scientific information is important for scientific progress but also carries risks of leakage of intellectual property and other technical information.[12] This is a particular concern for countries that the United States has designated as strategic competitors that have prioritized the acquisition of U.S. technical expertise. One proxy for this risk of technology leakage is **Metric I.G.1**, the percentage of publications coauthored with a researcher from a strategic competitor nation. A particular concern is research institutions affiliated with competitor nations' militaries, who may be studying military applications of these technologies. This risk is measured by **Metric I.G.2**, the number of unique U.S.-based authors who have coauthored at least one publication with an author affiliated with a strategic competitor's military-affiliated university. (We are not alleging that any of these authors are deliberately or accidentally leaking technical information to a foreign military. We are simply pointing out that these collaborations can be one possible source of intellectual property leakage that policymakers have identified as an area of concern and may want to monitor.[13])

II. Government Activity Metrics

National governments are typically the primary source of investment in basic science research and early-stage R&D, especially by academia, while industry focuses on later-stage R&D with a clearer path to eventual applications. Many quantum technologies are still at an early stage in technological maturity, so the degree and focus of government research funding is an important driver of academic progress in QIS. The following metrics attempt to quantify various key aspects of the governmental sector of the QIS ecosystem.

A. Overall Government R&D Investment

For a very early-stage technology like quantum, the government's primary role is generally funding R&D, and the simplest top-line summary statistic for government activity in the QIB is **Metric II.A.1**, the total amount of government funding in QIS R&D. In addition to

---

[11] The DoD's Basic Research Office lists "engaging with international partners" as one of its core goals in support of the department's overall mission.

[12] The skilled workforce ecosystem in QIS is highly international; and many scientific contributions come from international students, postdoctoral fellows, professors, and research scientists working in both academia and industry. Moreover, there is a shortage of talent both domestically and internationally. For a detailed discussion of the U.S. government's perspective on international talent in QIS in the context of balancing the promotion of innovation and the protection of national security, see National Science and Technology Council, *The Role of International Talent in Quantum Information Science*, October 2021.

[13] Torsten Oliver Salge, Erk Peter Piening, and Nils Foege, "Exploring the Dark Side of Innovation Collaboration: A Resource-Based Perspective," *Academy of Management Proceedings*, Vol. 2013, No. 1, 2013.

directly advancing the state of the art, a national government can shape its investment portfolio to ensure that a diversified portfolio of technologies is being pursued at the national level, as well as technologies that could advance strategic government priorities (e.g., national security). The government can also give R&D grants to support start-up companies as they develop new technologies—for example, through the Small Business Innovation Research (SBIR) and Small Business Technology Transfer (STTR) programs. Total R&D investment serves as a proxy for these activities.

B. Growth in Government R&D Investment

In addition to a point-in-time snapshot of government investment, it is important to know how that investment is changing over time. **Metric II.B.1** is the annual grown in total QIS R&D. Steady growth in investment indicates that QIS is a continued priority of policymakers and signals that a technologically leading nation is likely to sustain its leadership over time, which in turn could encourage private-sector activity.

C. Stability of Government R&D Investment

If government investment decreases before the private industry becomes self-sustaining, that could discourage new private-sector investment in QIS and threaten the financial stability of both academic programs and private companies. One way to quantify investment stability is via **Metric II.C.1**, the number and size of major multiyear investment programs made at the government-wide level. These investments convey a sustained demand signal that allows academic and private institutions to make major multiyear investments in human and physical capital, which is important for an early-stage technology with high capital requirements but uncertain applications. These sustained investments reduce the probability of a sudden decrease in research investment similar to the "AI winters" of the early 1970s and 1990s—a prospect that some experts worry may threaten quantum technology as well in the near future.[14]

D. Breadth of Government R&D Investment Sources

National government R&D investment can be concentrated in a single agency or distributed across multiple agencies, and each approach has its advantages and disadvantages. A centralized approach reduces the risk of duplication and can allow for focused and long-term development of specific technologies. But a decentralized approach increases technology diversification, provides stability against a sudden decrease in investment based on agency-specific contingencies (such as changes in senior personnel), and increases the range of government priorities being considered. Given the strategic goals outlined above, we believe that a broad range of funding sources is on net a positive feature, given the current stage of quantum technology and the high uncertainty regarding which technology approaches and applications will prove most valuable.

---

[14] Elizabeth Gibney, "Quantum Gold Rush: The Private Funding Pouring into Quantum Start-Ups," *Nature News Feature*, October 2, 2019a.

We use two metrics to measure this aspect of the ecosystem. **Metric II.D.1** is the number of significant funding sources for published research. Similar to the Metric I.C.2 discussed above, **Metric II.D.2** uses the Herfindahl-Hirshman index (HHI) to quantify the concentration of scientific research funding across the various funding agencies. This metric ranges from 0 to 1, with a higher value indicating that a small number of agencies dominate the research funding.

III. Private Industry Metrics

The desired end state of the QIB is a healthy and private-sector industry with stable revenue streams, which produces integrated technologies at useful scales that can deliver new capabilities. The private industry sector of the QIB is the sector that will eventually determine whether quantum technology proves transformative in practice or remains a primarily scientific pursuit. But as we will discuss below, the existence of a private-sector industry in most precision quantum technologies (with the exception of atomic clocks) is a fairly recent phenomenon, and it is challenging to assess the status of such a new industry that is still in very rapid flux. The following metrics attempt to assess various aspects of the private industry sector of the QIB.

A. Number and Distribution of Quantum Industrial Base Firms

This set of metrics attempts to give a snapshot of the nation's current overall QIS industry at the level of aggregated statistics for the companies in QIB. We have identified several dimensions along which to categorize firms, for which we believe the aggregate company distributions can convey useful information to policymakers.

These aggregate statistics give coarse but broad information regarding the overall state of the industry. (They are complemented by other metrics in the next category, which consider specific cutting-edge technical achievements by individual companies.) It is therefore difficult to draw firm conclusions from any of these, but taken together, we believe that they provide a useful profile of the overall industry sector and can serve as rough proxies for more-precise aspects of the QIB that are relevant to policymakers but difficult to directly assess. There is no clear optimal or baseline distribution to compare against, so we do not attempt to characterize any particular distribution as "good" or "bad." We believe that the most insight can be derived from these metrics by periodically reassessing these overall industry statistics and tracking trends over time; these trends may be more useful than the individual snapshots.

1. **Metric III.A.1** is the total number of firms in the nation's QIB. There are many caveats to this quantity: the boundaries of the QIB are somewhat subjective, not all firms in the QIB are necessarily important players, firms differ drastically by size,[15] and firm start-ups and failures are strongly influenced by economic cycles and market

---

[15] Which is partly determined by nation-specific laws, regulations, and customs, making international comparisons challenging.

bubbles, etc. The number of firms at any given time is therefore difficult to interpret or to compare across nations. However, we believe that time trends in this figure can provide useful information, with steady and sustained growth over time reflecting a healthy economic sector. Moreover, the number of firms is a rough proxy for the number of distinct quantum technologies being pursued, which we believe is important given the high technology uncertainty in QIS. The specific technology applications that the firms are pursuing can indicate to policymakers whether specific strategic technology priorities are covered by the nation's QIB.

2. **Metric III.A.2** is the distribution of firms by size (i.e., employees). Every size of firm has its own advantages: smaller firms can be dynamic, innovative, and quick to adapt to changing circumstances but tend to have a smaller financial cushion and so can be vulnerable to decreases in demand. Larger firms tend to have more financial resources and can sometimes make larger and longer-term capital investments, and they can experience both economies and diseconomies of scale.

3. **Metric III.A.3** is the distribution of firms by age. A large number of new firms can indicate a rapidly growing sector and many new technologies coming into maturity; on the other hand, these firms may not have built up many financial reserves or had time to establish best practices or demonstrate long-term profitability.

4. **Metric III.A.4** is the distribution of firms by funding level (e.g., venture capital funding or revenue). Trends in business income over time send a strong signal about the financial stability of the QIB; a sudden decrease in funding could signal an incoming collapse of private-sector activity in QIS. This is a particular concern given the early stage of the commercial sector in quantum technology and the lack of clear use cases for most of the nascent technologies: if many companies are dependent on VC funding rather than operating revenue, then they may be financially vulnerable to a loss of investor interest in this highly uncertain technology.[16] Like the employee count, income can also serve as a proxy for a firm's importance in the QIB, so the income distribution can give policymakers information about the degree of market concentration and about which specific technologies and applications are experiencing robust commercial activity.

5. **Metric III.A.5** is the distribution of firms across the different levels of the production chain. By this, we mean the different degrees of large-scale system integration represented by the firm's product, from basic components (e.g., mirrors or optical fibers) all the way up through fully integrated systems for the end user (such as a functional quantum computer). In order to produce final systems at useful scale, a healthy industrial base should have firms at all levels of this production chain, as well as supporting services. Although the best binning categories to use will depend on data availability and the maturity of the QIB, Figure 2.1 gives an example categorization scheme for this metric.

---

[16] Gibney, 2019a.

FIGURE 2.1

Example Binning Categories for Metric III.A.5.

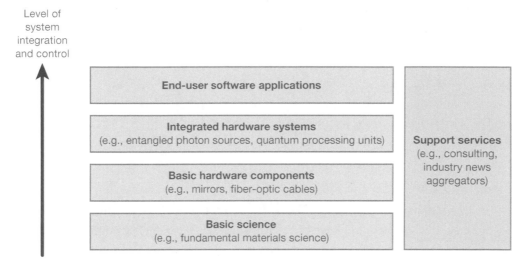

NOTE: This figure displays several of the key capabilities required for a mature quantum industrial base, roughly ordered from higher-level to lower-level systems. This is one possible way of categorizing the set of domestic QIB companies in Metric III.A.5.

B. Degree of Firm Specialization to Quantum Technology

Another industry-wide aggregate statistic that we believe to be useful is the categorization of QIB firms by degree of specialization in quantum technology. (This aspect is fairly distinct from the more traditional economic aspects described above, which is why we have broken it out separately.) Certain major firms in the QIB, such as the start-ups ColdQuanta, IonQ, Psi-Quanta, and Rigetti, are focused solely or primarily on quantum technology. Others, such as Google, Honeywell, and IBM, are large and established companies that produce products and services across a wide variety of technology areas. **Metric III.B.1** captures the proportion of firms in each category.

These two categories of firm each experience different risks to financial stability. Firms dedicated entirely to quantum technology are very unlikely to pivot entirely to another technology. But at this early stage of commercial deployment, they have little income beyond VC funding and government R&D grants, so they are vulnerable to failing or being bought if there is a future loss of commercial interest in these new technologies. On the other hand, established and profitable firms have more of a financial cushion to use corporate revenues to support longer-term and capital-intensive R&D programs that do not return an immediate profit. But they can shut down these programs at any time if they determine that there is too much uncertainly in the profit that they will eventually deliver. We therefore believe that a healthy industrial base should contain a balance of both categories of firm.

C. Foreign Supply-Chain Dependencies

In recent years, U.S. policymakers have becoming increasingly concerned about foreign supply-chain dependencies for critical technologies: one of President Biden's first actions upon taking office was to order an interdepartmental review of U.S. supply chains for these technologies.[17] Foreign dependencies create a risk of a loss of U.S. access to critical technologies, whether intentional or not. **Metric III.C.1** counts the number of critical quantum technology components or materials (as identified by SMEs) that can only be sourced from outside of the nation being assessed. **Metric III.C.2** separately identifies the subset of these dependencies that are on strategic-competitor nations, given the unique risks associated with these dependencies.

IV. Technical Metrics

This set of metrics takes a deeper dive into the specific technologies for which the nation is performing not only scientific research but also product development and commercial deployment. Some of the technical developments captured by these metrics have been achieved in academia and others in the private sector, so these metrics do not fit cleanly into the previous categories.

A. Innovation Potential

Patent metrics indicate the potential of U.S. industry and specific U.S. firms for innovation in quantum technologies. Patent filings describe new inventions that, if applied successfully to commercial applications, become innovations. The number and timing of quantum patent filings (**Metric IV.A.1**) relative to those of other countries indicate who is leading in advancing quantum technologies, and growth rate of filings (**Metric IV.A.3**) can indicate sustained progress. The diversity of filers, as quantified by the number of unique patent assignees (**Metric IV.A.2**), mirrors the diversity of the U.S. quantum portfolio. Patent filings also demonstrate access to key quantum technologies, including those technologies needed for integration at useful scale. Finally, quantum patent filings enhance the intellectual property portfolios of U.S. companies, increasing their long-term value regardless of market conditions.

B. Technical Achievement

This set of metrics captures the technical performance of the most advanced prototypes and devices produced within the nation. Unlike the other metrics listed in this report, the metrics in this category will not remain fixed over time; as quantum technologies become more mature, the relevant technical metrics (as determined by SMEs) will evolve to capture higher levels of integrated performance. But at a given point in time, they can be used to compare technical leadership in deployed technology.

---

[17] White House, *Building Resilient Supply Chains, Revitalizing American Manufacturing, and Fostering Broad-Based Growth: 100-Day Reviews Under Executive Order 14017,* June 2021.

In Chapters Three and Four, we have used the project team's expertise to select some of the key technical performance metrics for assessing the state of the art (as of 2021) in several important quantum technologies in each of the three application domains. We have applied those metrics to the most advanced devices described in the public technical literature in order to assess progress toward eventual applications. Quantifying the technical performance of prototype quantum systems is notoriously challenging—especially given the lack of immediate applications—and we do not claim that our selected metrics fully capture the relevant technical performance, but we believe that they give useful evidence for the current state of progress within various technologies. (We do not believe that these chosen metrics will still be the most relevant ones in several years, after the technologies have further matured.)

Unlike most of our other metrics, we chose not to separately study the U.S. and Chinese technical achievement. The technical state of the art naturally requires looking beyond one nation at a time, because the *global* state of the art is clearly a very important benchmark for comparison. We have therefore chosen several specific technologies in which the United States or China are arguably at or near the global forefront, and have listed the relevant technical metrics for these technologies in Chapters Three and Four together with other nations' demonstrated capabilities (where applicable).

C. Breadth of Technical Approaches in Development

Most quantum technologies are still at a low enough state of technology maturity that there are several basic physical approaches under pursuit. Moreover, different technical approaches can enable very different applications (especially in quantum sensing). Given the high uncertainty in the field as to which basic approaches will prove feasible and useful, we believe that a diversity of technical approaches and application goals increases the probability of eventual useful deployment.

Metric IV.C.1 is the number of distinct technology approaches under development or production in the nation (outside of basic scientific research). **Metric IV.C.2** counts the number of technology approaches in which the nation is the world leader, based on technical metrics such as those in Metric IV.B.1.[18] **Metric IV.C.3** is the number of technologies for which some firm (or other element of the industrial base) has announced a road map with concrete timelines to quantitative performance milestones that lead to eventual useful deployment. Although there is certainly no guarantee that these road maps will be met, publicly announced quantitative road maps do provide some accountability on the firm's part, indicate a sustained commitment to technology development over the long term, and give policymakers some (highly uncertain) estimates for when the technologies may become useful.

---

[18] The metrics in this class generally require more subjective judgment by SMEs than our other metrics, because there is inherently some degree of subjectivity as to where to draw the boundaries between distinct technologies. These judgments are somewhat subjective and may shift over time, but a large disparity between nations in this number indicates a qualitative difference in technology diversification. Similarly, SMEs must judge which metrics are the most important ones with which to determine world leadership.

Mapping the Metrics to the Strategic Goals

The discussion of each metric above attempts to tie in the aspect of the quantum ecosystem being captured with the strategic goals discussed earlier in the chapter. Table 2.2 summarizes this alignment. Each row corresponds to a set of (one or more) metrics, and each column corresponds to one of the strategic goals. An "x" in a given cell indicates that the metrics in that row can inform policymakers on progress toward the strategic goal in that column.

Overall Assessment Methodology

This report takes a mixed-method approach in order to make as holistic of an assessment as possible. Most of the research metrics are assessed through a big-data analysis of a large body of worldwide academic publications on quantum technology that was collected by a novel approach that combines automated keyword searching and SME judgment. The government metrics were mostly assessed from a review of government-issued reports and (for the China assessment) news media. The industry metrics were assessed by a combination of (a) individually profiling a representative sample of firms from their websites and financial reporting and (b) conversations with industry leaders. The technical metrics were mostly assessed from a review of the technical literature by SMEs on the research team. Appendix A discusses the methodological details.

By design, our metrics have different natural levels of scope. Some metrics attempt to describe the overall state of the nation's industry for a given application domain and so are applied to the national quantum sector in aggregate (or at least to a hopefully representative sample). Other metrics are more granular and attempt to capture the cutting edge in national capability in individual technologies and so necessarily have a more granular scope. We anticipate that different metrics will be more or less useful for different purposes, depending on policymakers' precise goals.

TABLE 2.2

Strategic Goals Addressed by Chosen Metrics

	Goals						
	I. Tech Leadership			II. Tech Availability			
	A. Current Tech Lead	B. Sustained Tech Lead	C. Diversified Tech Portfolio	A. Full Tech Production	B. No Competitor Dependencies	C. Government Priorities	III. Financial Stability
I. Research							
A. Overall research	X						
B. Research growth		X					
C. Institutional concentration	X		X				
D. Scientific impact	X						
E. Gov. priorities						X	
F. Institutional collaborations	X						X
F. Tech leakage					X		
II. Government							
A. Overall R&D investment	X		X			X	X
B. Investment growth		X					X
C. Investment stability		X				X	X
D. Breadth of investment sources			X			X	X
III. Industry							
A. Firm counts and sizes			X	X		X	X
B. Quantum specialization							X
C. Foreign dependencies				X	X		
IV. Technical							
A. Innovation potential	X	X	X	X	X		X
B. Technical achievement	X						X
C. Breadth of approaches	X		X	X			X

CHAPTER THREE

The United States' Quantum Industrial Base

We dedicated the majority of our research effort toward applying the metrics discussed in the previous chapter to the United States' QIB. This chapter summarizes our findings for each metric. Methodological details are provided in Appendix A. Wherever feasible, our top-line finding for each metric is either highlighted in bold font or set off in a stand-alone table. The state of the QIB is rapidly evolving, and we chose to use a uniform data-collection cutoff of July 2021 for all of our metrics.

Assessment of U.S. Quantum Information Science Research

The publication data used in the assessment below is drawn from the Core Collection of the Web of Science scientific publication database of over 90 million records from 21,000 peer-reviewed journals. Using a hybrid search strategy that iteratively combined automatic keyword searching and SME input, we collected a database of 46,016 worldwide academic publications in QIS published in well-regarded, peer-reviewed journals between 2011 and 2020 (inclusive). These publications were grouped into the three major QIS application domains based on keyword searches of the publication titles, author-provided keywords, and abstracts. See Appendix A for a detailed methodological discussion. We focus on presenting the U.S. results in this chapter but compare some of our findings to China's values as a point of contrast.

A. Overall Research Activity

Over the 2011–2020 period of analysis, the United States was the global leader in quantum computing publishing. During this period, authors from U.S.-based organizations published 7,319 scientific journal articles on quantum computing. Globally over the same period, 28,388 quantum computing articles were published, indicating that a U.S.-affiliated author was listed on approximately 26 percent of all quantum computing articles.

Over the 2011–2020 period of analysis, authors affiliated with U.S. organizations published 2,524 scientific journal articles on quantum communications. This indicates that a U.S.-affiliated author was listed on 15 percent of the 16,912 total quantum communications publications. During the same period, China was the global leader in terms of quantum communications publications. Authors affiliated with Chinese organizations published 6,440 scientific journal articles on quantum communications during the period.

Finally, U.S.-affiliated authors produced 1,240 quantum sensing journal articles. Given that there were 5,130 total quantum sensing publications over the period, this translates to a U.S.-affiliated author being listed on 24 percent of all quantum sensing publications. Table 3.1 provides the U.S. publication totals for the three QIS application domains. Table 3.2 provides the number of publications produced by the top ten publishing countries by application domain over the 2011–2020 period of analysis.[1]

TABLE 3.1

Total U.S. QIS Publications, 2011–2020 (Metric I.A.1)

Quantum Computing	Quantum Communications	Quantum Sensing
7,319	2,524	1,240

SOURCE: RAND analysis of Web of Science data.

TABLE 3.2

Total Publications by Highest-Publishing Countries, 2011–2020

	Quantum Computing Publications	Quantum Communications Publications	Quantum Sensing Publications
USA	7,319	2,524	1,240
China	7,050	6,440	1,539
Germany	2,749	1,258	648
Japan	2,275	936	334
UK	2,203	1,395	545
Canada	1,584	983	224
Italy	1,115	678	371
France	1,347	554	328
India	1,419	655	90
Russia	1,030	556	236

SOURCE: RAND analysis of Web of Science data.

B. Growth in Research Activity

Table 3.3 displays the compound annual growth rate (CAGR) for the United States for each of the three application domains. Figure 3.1 displays the annual number of publications from

---

[1] The ten included countries are the top ten publishing countries based on the sum of all three application domains.

TABLE 3.3

Compound Annual Growth Rate in U.S. QIS Publications, 2011–2019 (Metric I.B.1)

Quantum Computing	Quantum Communications	Quantum Sensing
10.8%	8.6%	8.2%

SOURCE: RAND analysis of Web of Science data.

FIGURE 3.1

U.S. QIS Publications by Year, 2011–2019

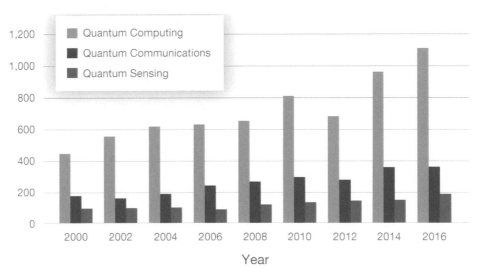

SOURCE: RAND analysis of Web of Science data.

2011 to 2019 for all three application domains for the United States.[2] The graph depicts a positive growth trend in all three application domains: the 2019 totals for all three application domains have more than doubled the 2011 totals.

C. Institutional Research Capacity

In the United States, over 2011–2020, 1,521 research units published at least one publication in one of the three application domains considered here. Table 3.4 provides the totals for each application domain. Table 3.5 depicts the top-20 U.S.-based research units by application domain.[3] Sixteen of the research units listed are the main campuses of large research universities. Two defense-focused national research organizations—Los Alamos National Laboratory

---

[2] Because the annual data for 2020 is not complete, we do not compute the annual growth rate between 2019 and 2020.

[3] The 20 included research units are the top-20 publishing research units based on the sum of all three application domains. Research units are sorted by quantum computing publications.

TABLE 3.4

Number of U.S. QIS Publishing Research Units, 2011–2020 (Metric I.C.1)

Quantum Computing	Quantum Communications	Quantum Sensing	Total
1,236	581	376	1,521

SOURCE: RAND analysis of Web of Science data.

TABLE 3.5

Top 20 U.S. Research Units by Application Domain, 2011–2020

Research Unit	Quantum Computing Publications	Quantum Communications Publications	Quantum Sensing Publications
MIT, Cambridge, Mass.	483	235	149
University of Maryland, College Park, Md.	399	111	48
Harvard University, Cambridge, Mass.	370	87	47
University of California Berkeley, Berkeley, Calif.	271	48	66
University of California Santa Barbara, Santa Barbara, Calif.	263	29	21
Caltech, Pasadena, Calif.	248	125	56
Princeton University, Princeton, N.J.	242	50	18
Stanford University, Stanford, Calif.	237	94	49
University of Michigan, Ann Arbor, Mich.	212	74	36
Yale University, New Haven, Conn.	205	57	23
Los Alamos National Laboratory, Los Alamos, N.M.	163	0	14
NIST, Boulder, Colo.	154	75	59
Purdue University, West Lafayette, Ind.	146	35	11
Oak Ridge National Laboratory, Oak Ridge, Tenn.	143	75	17
University of Colorado, Boulder, Colo.	137	32	66
University of Chicago, Chicago, Ill.	132	35	21
Texas A&M University, College Station, Tex.	121	63	30
NIST, Gaithersburg, Md.	97	67	37
Louisiana State University, Baton Rouge, La.	79	104	36
University of Rochester, Rochester, N.Y.	51	78	50

SOURCE: RAND analysis of Web of Science data.

and Oak Ridge National Laboratory—rank among the top-20 U.S.-based research units. The role of NIST in QIS publishing is worth highlighting; two NIST offices (Gaithersburg, Maryland, and Boulder, Colorado) play significant roles across the three QIS application domains.

The large contribution of universities to scientific research in QIS is evident when considering the full distribution of publishing across organization type. Table 3.6 depicts the total number of research units by organization type.[4] Table 3.7 depicts the share of U.S. publications in each application domain produced by a given organization type.

Although the lion's share of publications across the three application domains are produced by universities, the contributions of government research institutes and firms is not negligible. In total, government research institutes represent 13 percent of the affiliation slots across the three application domains. The top-five government research institutes in terms of total publication output are NIST, Boulder, Colorado; Oak Ridge National Laboratory, Oak Ridge, Tennessee; NIST, Gaithersburg, Maryland; Los Alamos National Laboratory, Los Alamos, New Mexico; and Sandia National Laboratories, Albuquerque, New Mexico.

In total, in the United States, firms occupied 8 percent of the QIS publication slots over the period. The top-five firms in terms of total publication output are Microsoft Research, Redmond, Washington; Raytheon, Cambridge, Massachusetts; IBM Thomas J. Watson Research

TABLE 3.6

Unique U.S. Research Units by Organization Type and Application Domain

	Academic	Corporate	Government
Quantum Computing	571	255	107
Quantum Communications	297	86	66
Quantum Sensing	206	51	54
Total	656	316	139

SOURCE: RAND analysis of Web of Science data.

TABLE 3.7

U.S. Publication Share by Organization Type and Application Domain

	Academic	Corporate	Government
Quantum Computing	78%	9%	12%
Quantum Communications	79%	7%	14%
Quantum Sensing	79%	5%	16%
Total	78%	8%	13%

SOURCE: RAND analysis of Web of Science data.

---

[4] Publications coauthored by two or more distinct organization types (e.g., an article published by a university and firm) are counted in both categories of organization. See Appendix A for details on how research units were categorized by organization type.

Center, Yorktown Heights, New York; IBM Corp, Yorktown Heights, New York; and Microsoft Research, Santa Barbara, California.

The HHI for research units measures how concentrated the production of publication output is across a country's research units. Table 3.8 depicts the HHI for the three major application domains as well as for QIS more generally. The concentration of U.S. research output is very low. For all three application domains, MIT is the research unit with the highest share of publication output. MIT is listed as an affiliation on 4.1 percent of quantum computing publications, 6.3 percent of quantum communications publications, and 7.6 percent of quantum sensing publications.

TABLE 3.8

U.S. Herfindahl–Hirschman Index for Research Units, 2011–2020 (Metric I.C.2)

Computing HHI	Communications HHI	Sensing HHI	Total QIS HHI
0.0101	0.0137	0.017	0.0106

SOURCE: RAND analysis of Web of Science data.

As discussed in Appendix A, we can gain some intuition for the meaning of the HHI by taking its reciprocal, which gives the equivalent number of firms in a hypothetical market with the same overall concentration but with all firms equally important. We find that the overall U.S. research concentration is the same if there were 94 distinct institutions that all published equally. This means that the U.S. QIS research sector has very low concentration, with a large number of roughly equally important research contributors.

D. Global Scientific Impact

Over the period of analysis, the United States produced 1,381 highly cited quantum computing publications. This is more than any other country and more than twice that of China (630) over the same period. During this period, there were a total of 3,043 highly cited quantum computing publications. This means that a U.S.-affiliated author was listed on 45 percent of highly cited quantum computing publications; a China-affiliated author was listed on roughly 21 percent of highly cited quantum computing publications.

Over the period of analysis, the United States produced 433 highly cited quantum communication. There was a total of 1,879 highly cited quantum communications publications. Thus, a U.S.-affiliated author was listed on 23 percent of highly cited quantum communications publications. Over the same period, China produced 577 highly cited quantum communications publications—translating to a Chinese-affiliated author being listed on roughly 31 percent of highly cited quantum communications publications.

From 2011 to 2020, the United States produced 235 highly cited quantum sensing publications, more than any other country. As there were a total of 570 highly cited quantum sensing publications, this means that a U.S.-affiliated author was listed on 41 percent of highly cited quantum sensing publications. Over the same period, Chinese-affiliated authors were listed on 148 highly cited quantum sensing publications, or 26 percent of the total.

Table 3.10 presents the number of U.S.-based research units to have produced a highly cited publication in each of the three QIS application domains. The total column refers to the number of unique U.S.-based research units to have produced at least one highly cited publication in any of the QIS application domains.

TABLE 3.9

Highly Cited U.S. QIS Publications, 2011–2020 (Metric I.D.1)

Quantum Computing	Quantum Communications	Quantum Sensing
1,381	433	235

SOURCE: RAND analysis of Web of Science data.

TABLE 3.10

Number of U.S. Research Units Producing Highly Cited Research, 2011–2020 (Metric I.D.2)

Quantum Computing	Quantum Communications	Quantum Sensing	Total
317	149	104	361

SOURCE: RAND analysis of Web of Science data.

E. Topical Alignment with Government Priorities

Based on publicly available reports and white papers released by DoD, we have identified three subdomains to be of low priority to DoD. As discussed in Chapter One, the Defense Science Board (DSB) has concluded that quantum key distribution "has not been implemented with sufficient capability or security to be deployed for DoD mission use,"[5] and the NSA has recommended against the use of QKD and quantum cryptography in National Security Systems.[6] We therefore designated QKD and quantum cryptography (very closely related applications) to be low-priority areas of quantum communications for DoD. Similarly, based on

TABLE 3.11

Percentage of U.S. Publications About Topics of Low Priority to DoD, 2011–2020 (Metric I.E.1)

Quantum Computing	Quantum Communications	Quantum Sensing
NA	27.6%	16.1%

SOURCE: RAND analysis of Web of Science data.

---

[5] Defense Science Board, 2019.

[6] National Security Agency, undated.

the DSB's conclusion that "quantum radar will not provide upgraded capability to DoD," we designated quantum radar and the related quantum illumination technologies to be a low-priority area of quantum sensing.[7] Both of these assessments are supported by the figure in the FY 2020 Industrial Capabilities Report to Congress that was adapted to Figure 1.1 in Chapter One.

We filtered the QIS publications for keywords that our SMEs judged corresponded to these low-priority areas. Over the 2010–2020 period of analysis, 27.6 percent of U.S. quantum communications publications focused on either QKD or quantum cryptography. Over the same period, 16.1 percent of quantum sensing publications were on the topic of quantum imaging. Figure 3.2 depicts the total publications for the United States across the three major application domains distinguishing for the subdomains (shaded in light blue and light orange) of low DoD priority.[8]

FIGURE 3.2

Total U.S. Publications by Application Domain with Low-DoD-Priority Subdomains

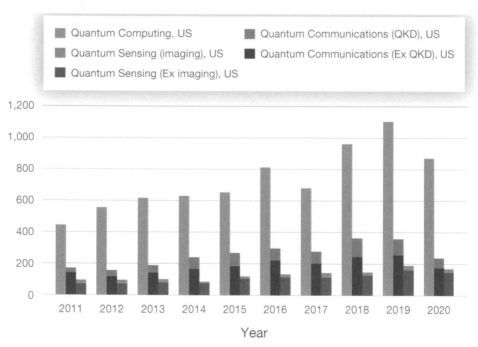

SOURCE: RAND analysis of Web of Science data.

---

[7] We did *not* include electromagnetic sensors such as Rydberg atom sensors among the low-priority applications. See Appendix A for the complete list of included keywords.

[8] The dip across all three application domains observed in 2020 is due to data incompleteness for this year.

F. Degree of Domestic and International Collaboration

Table 3.12 presents the average number of collaborating domestic institutions per domestic institution (i.e., network degree) for the three QIS application domains in the United States. Comparing these metrics to those of China (see Chapter Four for details) shows that U.S.-based research units collaborate more than their Chinese counterparts in quantum computing but less in quantum communications and quantum sensing.

Table 3.13 presents the share of U.S. publications that have an author based in another country. In all three application domains, the rate of international collaboration is higher among U.S. research units than their Chinese counterparts.[9] For all three application domains, China is the United States' top collaborating country. In the quantum computing application domain, Germany and Canada are the United States' second and third most frequent collaborators, respectively. In the quantum communications domain, Canada and the United Kingdom are the United States' second and third most frequent collaborators, respectively. For quantum sensing, the United States' second and third most frequent collaborators are Germany and the United Kingdom, respectively.

Figures 3.3–3.5 depict the collaborative networks of the top-20 U.S. research units for quantum computing, quantum communications (with QKD and quantum cryptography publications removed), and quantum sensing, respectively.[10] Research units based in the United States are shaded in blue, those based in China are shaded in red, and those based in other countries are shaded in green.

TABLE 3.12

Average Number of U.S. Domestic Collaborating Institutions, 2011–2020 (Metric I.F.1)

Quantum Computing	Quantum Communications	Quantum Sensing
11.37	6.07	5.90

SOURCE: RAND analysis of Web of Science data.

TABLE 3.13

Percentage of U.S. Publications Coauthored with Other Nations, 2011–2020 (Metric I.F.2)

Quantum Computing	Quantum Communications	Quantum Sensing
48.3%	51.9%	45.6%

SOURCE: RAND analysis of Web of Science data.

---

[9] It is worth noting the United States' largest funder of scientific research, the NSF, tends to encourage collaboration via its funding channels.

[10] Network plots made using Gephi. Nodes are sized proposal to publication counts, and edges are sized based on the collaborations between the two nodes. The network graph shows the top-20 U.S. organizations as well as the research units with whom they collaborate that have 50 or more quantum computing publications during the 2011–2020 period.

FIGURE 3.3

Collaborative Network for Top 20 US Publishing Organizations, Quantum Computing, 2011–2020

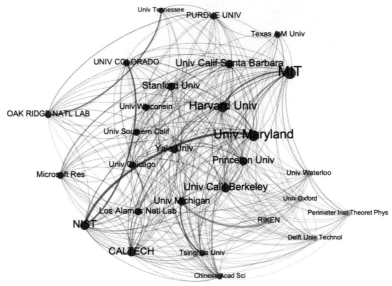

SOURCE: RAND analysis of Web of Science data.

FIGURE 3.4

Collaborative Network for Top 20 US Publishing Organizations, Quantum Communications, 2011–2020

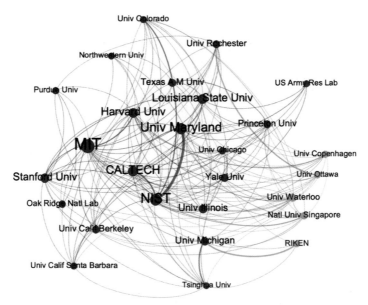

SOURCE: RAND analysis of Web of Science data.

NOTES: Network graph shows the top 20 US organizations as well as the research units with whom they collaborate that have 20 or more quantum computing publications during the 2011–2020 period. We exclude publications in the lower-priority subdomains of QKD and quantum cryptography.

FIGURE 3.5

Collaborative Network for Top 20 US Publishing Organizations, Quantum Sensing, 2011–2020

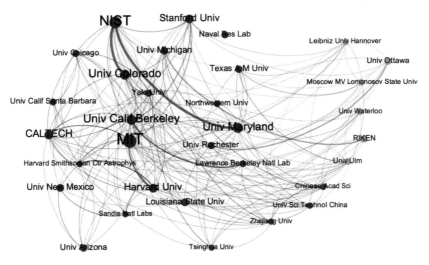

SOURCE: RAND analysis of Web of Science data.

NOTE: Network graph shows the top 20 US organizations as well as the research units with whom they collaborate that have ten or more quantum computing publications during the 2011–2020 period.

G. Risk of Technology Leakage

The publicly available summary of the 2018 National Defense Strategy of the United States of America states that "long-term strategic competitions with China and Russia are the principal priorities for the Department [of Defense]."[11] We have therefore identified China and Russia as the strategic competitor nations that are included in this class of metrics for this assessment (although the included nations might change in future years as the strategic environment shifts). Across all three application domains, collaboration with China was more common than with Russia by a factor ranging from four to ten depending on application domain. The collaboration of U.S.-based research units with China-based research units is visible in Figures 3.3–3.5 above (to ensure the interpretability of the graph, these figures depict only a subset of collaborations with strategic competitors). It is worth noting that Tsinghua University (one of China's preeminent research universities) is a top collaborator in all three network graphs.

Table 3.15 depicts the number of U.S.-affiliated authors that appeared on at least one publication with an author affiliated with a Chinese military university.

---

[11] Jim Mattis, *Summary of the National Defense Strategy of the United States of America: Sharpening the American Military's Competitive Edge*, Washington, D.C.: United States Department of Defense, 2018.

TABLE 3.14

Percentage of U.S. Publications Coauthored with Strategic Competitors, 2011–2020 (Metric I.G.1)

Quantum Computing	Quantum Communications	Quantum Sensing
13.0%	14.4%	14.2%

SOURCE: RAND analysis of Web of Science data.

TABLE 3.15

Number of U.S.-Based Authors to Have Collaborated with a Military-Affiliated University, 2011–2020 (Metric I.G.2)

Quantum Computing	Quantum Communications	Quantum Sensing	Total
5	2	0	6

SOURCE: RAND analysis of Web of Science data.

Assessment of U.S. Government Support

A. Overall Government R&D Investment

As discussed in the previous section, U.S. funding for QIS R&D is distributed across several different federal departments, which makes the total investment difficult to determine. The National Quantum Initiative Act resulted in the creation of the National Science and Technology Council's Subcommittee on Quantum Information Science, which has tracked total U.S. federal government QIS R&D investment since fiscal year (FY) 2019. Actual and estimated values for the total QIS R&D spending over FYs 2019–2021 as of January 2021 are displayed in Figure 3.6. **Actual spending in FY 2019 was $450 million, estimated spending in FY 2020 was $580 million, and the president's budget request for FY 2021 was $710 million.**[12] The figure breaks out how much of this spending was added to the prior baseline by the National Quantum Initiative, and how it was distributed across application domains. By FY 2021, the National Quantum Initiative was set to double QIS investments relative to the baseline.

---

[12] By comparison, the FY 2020 budget request contained $4.9 billion in basic and applied R&D spending on artificial intelligence (AI), another government technology priority. An important caveat to this comparison is that AI is at a higher level of technological maturity than quantum technology and is already being extensively deployed for multiple applications, so a higher level of spending is to be expected. Chris Cornillie, "Finding Artificial Intelligence Money in the Fiscal 2020 Budget," *Bloomberg Government*, March 28, 2019.

FIGURE 3.6

Total U.S. Federal Government Investment in QIS R&D, FY2019–2021

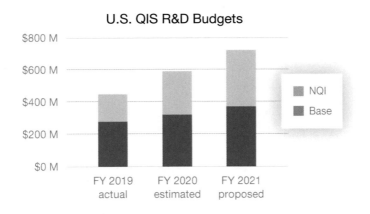

SOURCE: Subcommittee on Quantum Information Science, Committee on Science of the National Science & Technology Council, 2021.

NOTE: The top panel breaks out how much of the spending was authorized by the National Quantum Initiative. The bottom panel breaks the spending down by application domain. "QSENS" refers to quantum sensing, "QCOMP" to quantum computing, "QNET" to entanglement-based quantum communications and networking, "QADV" to basic research, and "QT" to new applications and supporting technology.

B. Growth in Government QIS R&D Investment

Since total federal investment in QIS R&D has been systematically tracked for only a few years, we do not have very much data with which to track time trends. However, the few data points that we have indicate **strong annual growth of 29 percent in FY 2021 and proposed 22 percent growth in FY 2022**. Most of this growth comes from the ramp-up of the National Quantum Initiative, although the baseline R&D budget is steadily increasing as well.

C. Stability of Government QIS R&D Investment

The National Quantum Initiative Act authorized **eight sustained multiyear funding initiatives** in QIS.[13]

NSF has been authorized to invest $75 million over five years to establish three Quantum Leap Challenge Institutes that are each led by a different university:

- Institute for Enhanced Sensing and Distribution Using Correlated Quantum States: Focusing on the development of quantum sensors for measurement
- Institute for Hybrid Quantum Architectures and Networks: Developing quantum processor networks
- Institute for Present and Future Quantum Computing: Advancing quantum computers and related algorithms.

DOE has been authorized to invest $625 million over five years to establish five QIS research institutes that are each led by a different national laboratory:

- Next Generation Quantum Science and Engineering: Fostering a supportive environment for innovation and commercialization of quantum technologies
- Co-design Center for Quantum Advantage: Advancing technical capabilities of quantum computers
- Superconducting Quantum Materials and Systems Center: Addressing decoherence issues and supporting the general advancement of quantum technology
- Quantum Systems Accelerator: Exploring the use of different quantum technologies and algorithms for various applications
- The Quantum Science Center: Supporting the quantum workforce and fostering development of revolutionary technologies by addressing technical challenges.

The other funding agencies each administer many smaller individual research programs, most of which last several years, but these programs are not codified in national legislation.

D. Breadth of Investment Sources

Table 3.16 depicts the number of distinct funding sources to have funded at least 50 publications authored by U.S. organization–affiliated authors.

Table 3.17 depicts the HHI for funding sources in the United States. Across all three application domains, the HHI for funding sources in the United States is 0.141. This is substantially lower than that of China (0.273), indicating that funding of QIS in the United States is less concentrated than in China. The HHI for funding sources in the United States is also

---

[13] After our July 2021 data collection cutoff, NSF announced the creation of two additional Quantum Leap Challenge Institutes, bringing the total up to ten: the Institute for Robust Quantum Simulation and the Institute for Quantum Sensing in Biophysics and Bioengineering.

TABLE 3.16

Number of Distinct Significant U.S. QIS Research Funding Sources, 2011–2020 (Metric II.D.1)

Quantum Computing	Quantum Communications	Quantum Sensing	Total
19	10	7	28

SOURCE: RAND analysis of Web of Science data.

TABLE 3.17

HHI for U.S. Funding Sources, 2011–2020 (Metric II.D.2)

Quantum Computing	Quantum Communications	Quantum Sensing	Total
0.151	0.131	0.135	0.141

SOURCE: RAND analysis of Web of Science data.

lower than that of China in all three application domains. As discussed in Appendix A, this concentration is equivalent to a hypothetical situation with seven equally important funders, indicating a diversified base of funding sources.

Federal agencies are responsible for funding the lion's share of the United States' quantum information publications. Eleven of the top-20 funders are federal government agencies. In fact, for each of the three application domains, the top funding agency for the United States is NSF. It is worth noting that intelligence- and defense-focused organizations are responsible for funding a large portion of the United States' QIS publications; six (U.S. Army, U.S. Air Force, DARPA, IARPA, U.S. Navy, DoD [no other specified]) of the top-20 funding organizations have organizational mandates centered on national defense and intelligence. Table 3.18 depicts the number of publications funded by top-20 QIS funding agencies in the United States.

The role of foundations in funding quantum information publications in the United States is noteworthy. Of the top-20 most productive funders, eight (John Templeton Foundation, Gordon and Betty Moore Foundation, Packard Foundation, Alfred P. Sloan Foundation, Simons Foundation, Foundational Questions Institute, The Welch Foundation, and W. M. Keck Foundation) are nonprofit foundations.

Assessment of U.S. Private Industry

A. Number and Distribution of Quantum Industrial Base Firms

The total number of firms that comprise the United States' QIB is very difficult to determine, as there is no comprehensive list and the criteria for inclusion are somewhat subjective. To make a very rough count, we combined several lists of key commercial stakeholders in quantum technology as of late 2020 (described in Appendix A) and created a list of **182 unique firms in the QIB as of late 2020** (Metric III.A.1). We were not able to investigate each firm in

TABLE 3.18

Number of Publications Funded by Funder, United States, 2011–2020

Funder	Quantum Computing Publications Funded	Quantum Communications Publications Funded	Quantum Sensing Publications Funded
NSF	2,741	713	381
U.S. Army	1,505	454	234
DoE	1,271	173	115
U.S. Air Force	893	340	173
IARPA	672	68	42
DARPA	477	251	281
U.S. Navy	354	238	61
John Templeton Foundation	239	178	39
Gordon and Betty Moore Foundation	185	47	25
Alfred P. Sloan Foundation	142	32	21
NIST	136	52	65
Microsoft Research	122	6	1
Packard Foundation	120	45	32
DoD (no other specified)	105	10	16
Simons Foundation	92	38	12
NASA	87	42	40
National Institutes of Health	71	1	28
Foundational Questions Institute	47	61	23
The Welch Foundation	45	5	8
W. M. Keck Foundation	33	10	13

SOURCE: RAND analysis of Web of Science data.

detail and determine its importance in the QIB. This count probably missed some firms and may have included a few that are only tangentially relevant to the QIB, but we believe that is a reasonably accurate count of the relevant commercial players in quantum technology as of late 2020. The commercial sector in quantum technology is rapidly changing, so this number may have changed significantly since then.

For the sake of repeatability and transparency of methodology, for the other metrics in this category we specialized to one data set, the list of companies in the Quantum Economic Development Consortium (QED-C), the main U.S. quantum industry-wide stakeholder consortium, which was established by the National Quantum Initiative under operation by NIST. These 139 companies (as of September 2020) covered the majority of our list (including all of the companies that our SMEs judged to be the most important players), and although they do not cover the entire industrial sector, we believe them to be a fairly representative sample. We individually profiled each member company along each of the dimensions captured by these metrics as best we could based on publicly available information (which was often quite sparse).

Figure 3.7 shows the number of companies involved in each application domain. Several companies produced products or services that we judged applied across all three domains, and we characterized those as "cross-cutting." Nearly half of the QED-C companies worked on quantum computing, and a large proportion made cross-cutting products and services (often basic hardware components). Despite the large amount of U.S. publication activity around quantum communications, very few companies focused on that domain.

The rest of the company distributions reported in this section cover only the 50 QED-C companies (all start-ups as of our data collection cutoff[14]) that we categorized as primarily or exclusively focused on quantum technology, because the overall characterization of the large diversified firms is less relevant to the QIB, and we were unable to determine the sizes of their dedicated quantum programs. Given the small sample sizes, we did not further break down the companies by application domain.

FIGURE 3.7

Distribution of QED-C Companies by Quantum Application Domain

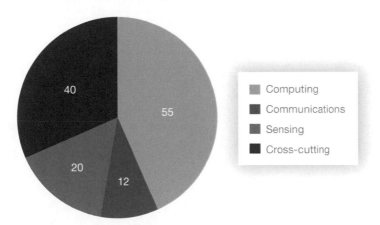

NOTE: Companies whose products fell into two application domains were counted in both; companies whose products fell into all three were categorized as "cross-cutting." We were unable to categorize 12 companies based on public information.

---

[14] Two companies dedicated exclusively to quantum technology (IonQ and Rigetti Computing) went public or announced plans to do so after our July 2021 data collection cutoff. As publicly traded companies, these are generally no longer considered to be start-ups. Mark Sullivan, "How IonQ Is Planning to Bring a Quantum Computer to the Masses," *Fast Company*, October 1, 2021; Frederic Lardinois, "Rigetti Computing Goes Public via SPAC Merger," *TechCrunch*, October 6, 2021.

Figure 3.8 shows the distribution of employee counts across the 29 quantum-focused QED-C firms that reported this information (Metric III.A.2). The large majority of these companies were fairly small, with only five companies having over 50 employees.

Figure 3.9 shows the distribution of founding years across the 32 quantum-focused QED-C companies that reported this information (Metric III.A.3). Many of these companies are quite young, with most of them having been founded since 2017. The rapid growth indicates high private-sector interest in quantum technology, but the many start-ups that are still very new have not necessarily demonstrated a stable business model, product line, or revenue stream, and they may be economically vulnerable to downturns in the quantum technology sector.

FIGURE 3.8

Distribution of Quantum-Focused QED-C Companies by Employee Count

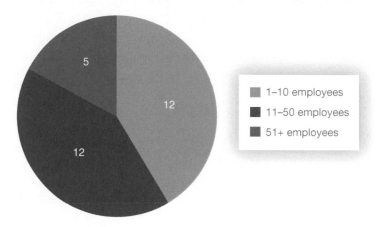

FIGURE 3.9

Distribution of Quantum-Focused QED-C Companies by Founding Year

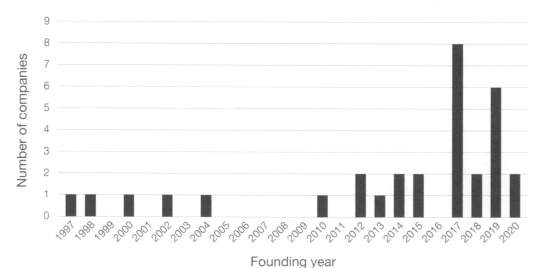

NOTE: One company founded before 1997 is omitted for space.

Figure 3.10 shows the distribution of total announced VC funding across the 20 quantum-focused QED-C companies that provide this information publicly (Metric III.A.4). Our primary data source was the Crunchbase platform; Appendix A describes our methodology. As of June 2021, a total of $1.28 billion in VC funding had been announced across these 20 firms, but more than three-fourths of that total went to just three firms, all of which are in quantum computing: PsiQuantum ($509 million), D-Wave Government Systems ($256 million), and Rigetti Computing ($199 million).[15]

In order to give additional context to this distribution, Figure 3.11 compares it to the approximate distribution of VC raised across a large sample of over 1,000 technology start-ups across multiple sectors.[16] We made several assumptions in constructing this distribution, which are explained in Appendix A.[17]

We see that the distribution of VC funding across quantum-focused firms is much more top-heavy than the distribution across the entire technology sector, with just a few huge companies dominating the VC funding. This indicates that the VC market is placing high confidence in a few major quantum start-ups, which indicates that these firms have a quite stable

FIGURE 3.10

Distribution of Quantum-Focused QED-C Companies by Announced Venture Capital Funding

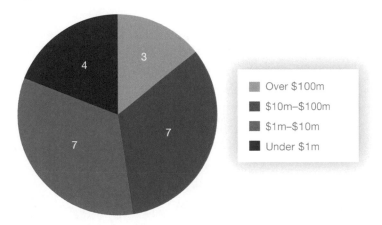

SOURCE: RAND analysis of Crunchbase data.

---

[15] Sixteen quantum computing companies had announced a total of $1.27 billion in VC funding, six quantum communications companies had announced a total of $25 million, and five quantum sensing companies had announced a total of $107 million. The 13 highest-funded companies were all in quantum computing. Some companies work in multiple domains and were counted multiple times. None of the top 15 investors in quantum computing have specific ties to national security, although In-Q-Tel has made more modest investments in Rigetti.

[16] CB Insights, "Venture Capital Funnel Shows Odds of Becoming a Unicorn Are About 1%," *CB Insights.com*, September 6, 2018.

[17] As discussed in the Appendix, constraints on data availability forced us to use unevenly sized bins.

FIGURE 3.11

Distributions of All Tech Start-Ups and of Quantum-Focused Start-Ups by Venture Capital Funding

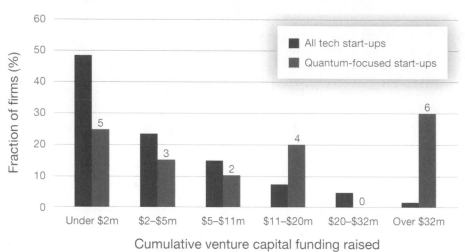

SOURCE: RAND analysis of Crunchbase and CB Insights data.
NOTE: The absolute count of quantum-focused QED-C start-ups within each bin are displayed above the orange bars.

financial base; however, if they fail for any reason, a large portion of the private investment in quantum technology will be lost.

Figure 3.12 shows the distribution of most QED-C companies by the broad category of product or service that they provide (Metric III.A.5),[18] using the categories illustrated in Figure 2.1. There is a fairly uniform distribution across our chosen categories, with the majority of companies producing hardware, split roughly evenly between basic components and integrated systems. A more complete analysis would need to individually assess the firms within each category, but there are no obvious gaps in the chain of production.

B. Degree of Firm Specialization in Quantum Technology

Figure 3.13 shows the distribution of QED-C companies by their degree of specialization to quantum technology. The companies in the first category are start-ups dedicated to quantum technology. Those in the second category are not dedicated to quantum technology, but either (a) they produce a commercial product that we judged to be relevant to quantum technology, or (b) they report a dedicated internal quantum R&D program. For the companies in the third category, we were unable to find evidence of a dedicated product or research program. We believe that most of the companies in this category are monitoring the develop-

---

[18] We omitted the companies that did not clearly have a dedicated quantum technology program, as discussed in the next subsection. Some companies were assigned to multiple categories.

FIGURE 3.12

Distributions of QED-C Firms by Primary Product or Service Provided

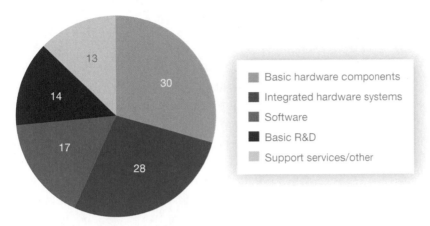

FIGURE 3.13

Distributions of QED-C Firms by Degree of Specialization to Quantum Technology

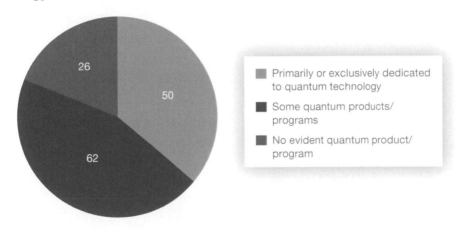

ment of the field for potentially useful applications but are not actively advancing the field themselves. However, it is possible that they are funding internal research that they have not publicly announced. We see that there is a relatively even balance of firms that are partially versus fully specialized to quantum technology.

C. Foreign Supply-Chain Dependencies

In order to analyze the supply chains for key components in quantum computing, communications, and sensing, we spoke with representatives of the following organizations: Cold-Quanta, FieldLine, Google, IBM, IonQ, MIT Lincoln Laboratory, Qunnect, Twinleaf, and

Vector Atomic.[19] Five of these organizations are working on quantum computing, one on quantum communications, and three on quantum sensing. Our research questions and analysis methodology are summarized in Appendix A.

Dependence on Foreign Suppliers

The organizations we spoke with source components from foreign suppliers in the following countries: Germany, Finland, United Kingdom, the Netherlands, Sweden, France, Italy, Russia, Canada, China, Japan, Australia, Philippines, Taiwan, and Malaysia. The primary components provided by these countries are lasers and electronics. In Europe, there are three companies that were highlighted by multiple organizations: TOPTICA Photonics (and subsidiary TOPTICA Eagleyard) in Germany, which provides laser diodes; Low Noise Factory in Sweden, which provides high-electron mobility transistor amplifiers; and Bluefors in Finland, which provides dilution refrigerators. In general, Germany was highlighted as an important source of high-quality lasers. Table 3.19 provides an overview of components sourced from Europe by the organizations we spoke with.

TABLE 3.19

Quantum Technology Components Sourced from European Suppliers

Component	Country (Supplier)
Single-photon detectors	Germany, Italy, France, Sweden
Laser diodes	Germany (Toptica Photonics)
Microcontrollers	Italy, France
Dilution refrigerators	Finland (Bluefors), United Kingdom (Oxford Instruments), the Netherlands (Leiden Cryogenics)
High Electron Mobility Transistor Amplifiers	Sweden (Low Noise Factory)
Optical lithography tools	The Netherlands (ASML)
Dielectric Glass Windows for vacuum chambers	Germany (Schott)
Fiber Phase Modulators	France
Double-angle evaporators	France
200mm sapphire wafers	Russia
Monolithic integrated windows for vacuum chambers	United Kingdom (UK Atomic Energy Authority)
Entanglement sources	Unnamed European countries

SOURCE: RAND analysis of conversations with industry representatives.

---

[19] We agreed not to identify any individuals by name, not to ascribe any specific statements to any individual organization, and not to discuss any proprietary information. No compensation was exchanged between RAND and any of these organizations. These organizations did not review this report before its release, and their agreement to speak with us should not be interpreted as constituting an endorsement of any of its findings.

The other primary supplier region is Asia, where Japan and China stood out as important supplier countries for the U.S. QIB. Japan was mentioned by multiple organizations because of the company Nichia, which was described as the primary global supplier for blue gallium nitride diodes used in blue lasers. The components and materials supplied by China tend to be less specialized—for example, commercial off-the-shelf (COTS) components such as electronics and optics. U.S. firms tend to buy components from China because of their low price and not because Chinese firms provide unique advanced technology components. Table 3.20 provides an overview of components sourced from Asia.

Additional foreign dependencies include lasers provided by MOGLabs in Australia and potentially by some Canadian companies, as well as rubidium-87, quarter-wave plates, and lenses, which had unknown or unspecified foreign sources. In some cases, a QIB firm may purchase a material or component from a U.S. distributor who may not tell the firm where they source the material or component.

As shown by Tables 3.19 and 3.20, most specialized components obtained from foreign sources come from countries allied with the United States; however, China stands out as a key supplier for COTS components and certain raw materials. Based on our conversations, China is the primary supplier for these components due to significant differences in cost between Chinese components and those sourced elsewhere. For example, printed circuit boards and mirrors for quantum computing applications can be an order of magnitude more expensive if sourced domestically. Russia provides fewer components for quantum applications; however,

TABLE 3.20

Quantum Technology Components Sourced from Asian Suppliers

Component	Country (Supplier)
Microcontrollers	Philippines, Taiwan, Malaysia, China
COTS electronics	China
COTS digital-to-analog converters	China
COTS analog-to-digital converters	China
COTS optics and raw materials for optics	China
Nonlinear crystals	China
Nonlinear optics	Japan
Mirrors	China
Blue gallium nitride laser diodes	Japan (Nichia)
Cables	Japan (COAX CO.)
200mm sapphire wafers	Japan (Kyocera)
Electron beam lithography tools	Japan
Distributed Bragg reflector laser diodes	Unnamed Asian countries

SOURCE: RAND analysis of conversations with industry representatives.

there is only one non-Russian company that produces large sapphire wafers used in some trapped ion quantum computing systems. Sapphire wafers can also be used in the manufacture of superconducting-transmon-based quantum processors.

Reasons for Foreign Dependencies

The organizations we spoke with provided multiple explanations for foreign dependencies in their supply chains, including relative component costs, differences in quality, lack of domestic alternatives, and acquisition of domestic suppliers by foreign companies. They also noted that lack of visibility into lower-tier suppliers likely obscures some foreign dependencies.

As previously noted, low production costs in China make sourcing COTS components from there economical, but U.S. quantum organizations also source single-photon detectors and entanglement sources from Europe for cost reasons. For both of these components, the European supply chain is both more extensive and more cost-competitive. European suppliers are also more capable of meeting the specifications for these components and are additionally known for producing the best high-electron mobility transistor amplifiers and dielectric vacuum chamber windows. At least one company also characterized previous work with a domestic laser supplier as a mistake due to resulting quality issues, and they suggested that an Asian supply chain may have been more reliable.

A lack of suppliers in the United States in specific areas leads to foreign supply-chain dependencies. As noted above, the domestic supply chain for single-photon detectors and entanglement sources is small (one or two companies) compared to the European supply chain. The dearth of companies providing single-photon detectors—used in quantum communications in the United States may be due in part to the relative lack of emphasis on quantum communications in the United States. There are also few or no domestic suppliers for high-electron mobility transistor amplifiers, cables for quantum computing, and double-angle evaporators, particularly when accounting for required quality and scale. However, in some cases, for example with double-angle evaporators, such components may not be needed to fabricate the quantum technology components, and so this foreign dependence may not represent a critical dependency.

Nevertheless, there are critical foreign dependencies in the U.S. QIB supply chain. This is partly caused by the acquisition of U.S.-based suppliers by foreign companies. For example, the high-quality laser diode supplier TOPTICA Eagleyard, which is now a subsidiary of Munich-based TOPTICA Photonics, was formerly a U.S. company. Another German company, OSRAM Opto Semiconductors, acquired the American company Vixar, which supplies lasers for quantum sensing applications.

The people we spoke with also noted that their organizations do not always have insight into the original origin of the components or materials they use. Even if they use a U.S.-based vendor, this does not guarantee that the vendor is not obtaining their products from a lower-tier supplier outside the United States. This lack of visibility was specifically noted for electronics, rare-earth magnets, and rubidium-87.

Dependence on Limited Supplier Base

Reliance on a limited set of suppliers is an additional source of risk for the U.S. quantum technology supply chain. This issue impacts lasers, optics, heat vapor cells, sapphire wafers, low-noise amplifiers, certain cables, double-angle evaporators, and various atomic isotopes. Having a limited set of suppliers increases the risk that changes to a supplier's catalog could cut off component supply completely and can force companies to work with unreliable components due to a lack of alternatives. There may be other system designs that could resolve issues that arise, but it can be difficult to make substantial changes once technology has been designed around a specific class of components.

For lasers, the issue of limited supply chains coincides with that of foreign dependencies. German companies (particularly TOPTICA Photonics) supply high-quality lasers with specifications other sources are unable to meet, and Nichia in Japan is the only source of blue gallium nitride lasers suitable for quantum applications. Rubidium lasers with required performance specifications can only be sourced from TOPTICA Eagleyard or a small number of Asian suppliers. Although production of crystals for lasers is not necessarily limited to a specific company, there are very few suppliers outside of China, although some laser manufacturers grow crystals in-house for their own products.

Many of the other components with limited supply chains are also sourced from foreign companies; 200mm sapphire wafers are difficult to obtain and are only produced by one Japanese company and a small number of Russian companies. Low Noise Factory in Sweden is the go-to supplier of low-noise amplifiers in the quantum field, double-angle evaporators are made by a single organization in France, and one company we spoke with only sources cables from a company in Japan. Even though there is a U.S. option for heat vapor cells, one company claimed that a German supplier produced a much higher-quality product, albeit a more expensive one.

Optics and atomic isotopes have limited supply chains, but there are either U.S. suppliers or they were not highlighted as having foreign suppliers in our conversations. Very few suppliers can meet the optical polishing specifications of one organization we spoke with. The atomic isotopes of concern in our conversations were calcium-43, barium-143, and isotopically pure ytterbium, all of which must be obtained from Oak Ridge National Laboratory. One company's representative said that it may be able to find other, potentially foreign, sources of ytterbium if it could no longer obtain it from Oak Ridge but that its current dependence on a single source is a source of concern.

Reasons for Limited Supply Chains

Quantum technology organizations provided a number for explanations for the lack of extensive options for certain components. In many cases, quality was the driving factor—other companies, even those manufacturing some version of the component, could not meet specification requirements. Characteristics of the quantum technology development process also sometimes hindered development of a robust supply chain. Because many quantum technology companies require components in relatively small numbers, it is not always economical for potential suppliers to produce those components. Quantum technology is also developing rapidly, and it can be expensive to search for multiple vendors for every component when the component requirements are not stable. Finally, certain components and materials can be made by only a handful of organizations. Blue gallium nitride lasers are difficult to make, making Nichia the unique supplier for laser diodes that operate at these frequencies. Although ytterbium is not particularly rare, producing isotopically enriched ytterbium requires highly specialized facilities found only in DOE labs. This material may also be available from nuclear facilities overseas but is not readily available as a commercial product.

Critical Components and Materials

In discussing critical supply-chain dependencies, the organizations we spoke with highlighted a handful of key components used in their systems. Lasers, which are used in multiple quantum applications, were the most commonly mentioned critical component. Frequency-stable lasers in the ultraviolet to mid-infrared frequencies are critical for multiple quantum computing and sensing companies. This is of particular note given the extent to which the supply chain for certain high-quality lasers is concentrated in a relatively small number of foreign suppliers. In additional to lasers, field-programmable gate array microchips are important for generating control signals in quantum computing. Components and materials associated with dilution refrigerators are also critical, including electronics that can operate at low temperatures and can be placed inside the dilution refrigerators as well as helium-3, which is needed to obtain millikelvin temperatures. For trapped ion computing, critical components and materials include the atoms used to make the qubits, such as calcium-43 and barium-143, connectors, electrodes, and vacuum chambers. Although lasers were the most frequently discussed critical quantum sensing component, rubidium-87 isotopes are also important for certain applications.

Other Relevant Insights

As touched on previously, the small scale of quantum technology development can make acquiring certain components difficult. Large companies with low margins are sometimes uninterested in producing dozens of components, particularly if it involves changes to their manufacturing process. This can prevent companies from finding new suppliers as well as jeopardize existing suppliers. One person we spoke with described losing a supplier because they were bought out by a large company that shifted their focus to a larger consumer market and then declined to make custom products. Because these problems can cause small quantum companies to turn to small suppliers for low-volume components, the supply chain gains additional vulnerability. Disruptions to the already low-volume and intermittent demand for specialized quantum components could cause financial troubles for small suppliers living contract to contract, leaving them open to foreign acquisition.

To some extent, this issue is driven by the nature of quantum technology and the R&D process. In quantum computing, for example, the end goal for a company may be to develop a relatively small number of high-performance quantum computers, instead of thousands of mass-market devices. Although this leads to the sourcing challenges already mentioned, it also means potential scarcity of raw materials is less of an issue because those materials are only required in small amounts.

Another inherent characteristic of certain quantum technologies is the degree to which the atoms used drives the laser wavelengths required. Because laser options are constrained by nature, companies face trade-offs between (a) choosing a potentially less-than-ideal atom as a qubit but one that is close to a common wavelength so they can use lasers already developed for other applications, and (b) developing a new laser that operates at a frequency that is optimal for a qubit, which makes the firm reliant on a unique supplier specializing in lasers at more-niche wavelengths. The pace of technology change also poses problems for companies; because their technology is still evolving, it is difficult to know which components will be needed in the future, so supply-chain issues cannot be easily resolved by stockpiling components in large numbers.

Assessment of U.S. Technical Metrics

A. Innovation Potential

This subsection presents the results of an analysis of quantum technology patent data derived from the IFI CLAIMS Direct Platform global patent database. As described in more detail in Appendix A, we assessed our metrics for quantum computing, quantum communications, and quantum sensing, counting for each sector all U.S. patent applications containing any of the keywords used in the analysis of quantum publications.[20] Table 3.21 summarizes our top-level patenting metrics for all patents filed in the United States through 2019. Taken together, they demonstrate substantial innovation potential in all three sectors.

TABLE 3.21

U.S. Patenting Metrics in Quantum Technology Through 2019

Metric	Quantum Computing	Quantum Communications	Quantum Sensing
Total patent applications (Metric IV.A.1)	4,845	1,385	787
Number of unique patent assignees (Metric IV.A.2)	1,296	755	610
Annual growth in patent applications (Metric IV.A.3)	17%	12%	14%

SOURCE: RAND analysis of IFI CLAIMS Direct patent data.

NOTE: Annual growth in patent applications refers to the compound average growth rate in applications filed over the 2010–2019 period.

---

[20] A family of patent applications may be submitted on a single invention. We count each family only once and record the priority year (typically the year of first submittal). We also keep track of the submittal year of each application in the family.

Figure 3.14 shows the cumulative number of U.S. patents filed in each application domain for each year from 2000 to 2020. (Since a relatively small number of patents are filed each year, the annual counts are noisy, and cumulative counts more clearly demonstrate long-term trends.)

FIGURE 3.14

Cumulative U.S. Quantum Technology Patent Applications, 2000–2019

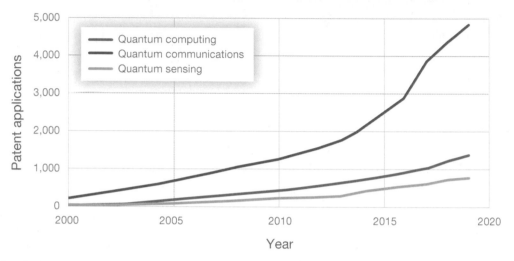

SOURCE: RAND analysis of IFI CLAIMS Direct patent data.

Quantum Computing

After a linear increase of about 100 applications per year between 2000 and 2010, cumulative U.S. patent applications began to increase exponentially—the start of the S-curve behavior typical of an emerging technology).[21] Since these (U.S.) S-curves typically take 18–20 years to reach saturation, we estimate that the current total of 4,845 is somewhere near the middle of the curve.[22] Comparing U.S. quantum computing patent applications to those of other countries, we find that the United States has the earliest emergence and the most applications—both signs of technological leadership. However, the closest competitor, China, has an emergence starting at a lower level but at about the same time and a more steeply rising S-curve, as described in the next chapter. China has also made significant recent advances in quantum computing, as described in the technical metrics section.

We also observed this S-curve behavior for cumulative patent applications in the subfield of superconducting quantum computing, defined by the keywords "superconductor"

---

[21] Christopher A. Eusebi and Richard Silberglitt, *Identification and Analysis of Technological Emergence Using Patent Classification*, Santa Monica, Calif.: RAND Corporation, RR-629-OSD, 2014. Growth from 2000 to 2010 is roughly linear.

[22] This estimate is based on statistical analysis of thousands of U.S. S-curves.

and "qubit." We show the cumulative number of U.S. patent applications for superconducting quantum computing in Figure 3.15, which constitutes almost one-third of total U.S. quantum computing patent applications and appears to be rising more sharply than Figure 3.14.

FIGURE 3.15

U.S. Superconducting Quantum Computing Patent Applications, 2000–2018

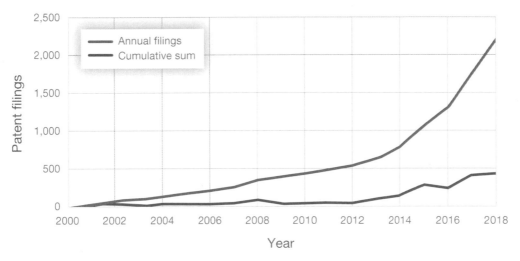

SOURCE: RAND analysis of IFI CLAIMS Direct patent data.

The U.S. patent applications shown in Figure 3.14 were assigned to 1,296 unique entities. Table 3.22 shows the ten entities with the most applications. These include large U.S. companies (Intel, IBM, Google, and Microsoft), D-Wave Systems (a Canadian company with U.S. offices), and large Japanese companies (Nuflare is a spin-off from Toshiba). Other filers include the Canadian company 1QB; U.S. companies Rigetti, IonQ, and PsiQuantum; several U.S. and foreign universities and government laboratories; and Chinese companies Huawei and Alibaba.

Quantum Communications

From the quantum communications curve in Figure 3.14, we see an S-curve emergence beginning in 2000, rising from ten applications in that year to over 1,300 in 2019. However, comparing U.S. quantum communications patents to those of other countries, we find that China has almost three times as many quantum communications patent applications as the United States and a much more steeply rising emergence S-curve that begins at almost the same time, indicating that China may be the technological leader in this area.[23] This is con-

---

[23] When interpreting Chinese patent totals, it is important to recognize that on average, Chinese patents have been founded to be of lower quality than U.S. patents and that patenting in China is often motivated by political, rather than purely commercial, incentives. Jon Schmid and Fei-Ling Wang, "Beyond National Innovation Systems: Incentives and China's Innovation Performance," *Journal of Contemporary China*, Vol. 26, No. 104, 2017.

TABLE 3.22

Assignees with the Largest Number of U.S. Quantum Computing Patent Applications

Assignee	Number of U.S. Quantum Computing Patent Applications
Intel	876
IBM	455
D-Wave Systems	257
Microsoft	152
Nuflare Technologies	113
Google	95
NEC	69
Toshiba	60
Fujitsu	50
Canon	42

SOURCE: RAND analysis of IFI CLAIMS Direct patent data.

NOTE: Includes all applications on which entity was an assignee, so applications may be counted more than once.

sistent with China's recent successes in sending quantum encrypted messages over long distances using its *Mozi* satellite.

The U.S. patent applications were assigned to 755 unique entities. Table 3.23 shows the ten entities with the most applications. These include large U.S. companies (IBM, Hewlett-Packard, AT&T, Procter & Gamble, Kodak), quantum-focused companies (MagiQ, D-Wave Systems), government and academic research institutions (U.S. Army Research Laboratory, MIT, NSF), and a VC firm (NextGen Partners). Other filers include a broad range of U.S. and foreign companies, universities, and laboratories.

Quantum Sensing

From the quantum sensing curve in Figure 3.14, after a very gradual rise through the early 2000s, we see an S-curve emergence beginning in 2005, rising from around 100 applications to almost 800 in 2019. Comparing U.S. quantum sensing patent applications to those of other countries, we see that the United States has the greatest number of applications and the earliest emergence. However, because of the diversity of types of quantum sensors, technical analysis of specific sensor types and applications is required to understand technological leadership in this sector.

The U.S. patent applications were assigned to 610 unique entities. The assignees are a diverse group, including large and small companies based in the United States and abroad, universities, and government laboratories. This diversity is reflected in Table 3.24, which shows the ten entities with the most applications.

TABLE 3.23

Assignees with the Largest Number of U.S. Quantum Communications Patent Applications

Assignee	Number of U.S. Quantum Communications Patent Applications
MagiQ	131
IBM	113
Hewlett-Packard	64
D-Wave Systems	46
NextGen Partners	45
U.S. Army Research Laboratory	40
AT&T	37
MIT	35
Procter & Gamble	33
Kodak	33
NSF	33

SOURCE: RAND analysis of IFI CLAIMS Direct patent data.

NOTE: Includes all applications on which entity was an assignee, so applications may be counted more than once.

TABLE 3.24

Assignees with the Largest Number of U.S. Quantum Sensing Patent Applications

Assignee	Number of U.S. Quantum Sensing Patent Applications
Microsoft	83
Equal1 Labs	28
Honeywell	23
MIT	19
NextGen Partners	18
Boeing	12
Caltech	12
Charles Stark Draper Laboratory	11
Qubitekk	11
Raytheon	11

SOURCE: RAND analysis of IFI CLAIMS Direct patent data.

NOTE: Includes all applications on which entity was an assignee, so applications may be counted more than once.

B. Technical Achievement

Quantum Computing

Quantum computers operate in fundamentally different ways than digital computers, so traditional technical performance metrics such as number of transistors in the central processing unit or microprocessor speeds have no meaning for quantum computers. Instead, we use technical metrics specific to quantum computing.

The technical achievement metrics for U.S. companies and academic institutions that are leading developers of quantum computers are shown in Table 3.25. U.S. organizations are pursuing the development of a wide range of technical approaches to quantum computing, and not all of these organizations and approaches are shown in the table but only the most mature prototypes whose technical performance has been publicly documented.

The two leading U.S. quantum computing developers are currently Google and IBM. They have developed systems with 53–65 qubits and have published papers describing their analytic and benchmark testing results. The more qubits available for computation, the more

TABLE 3.25

U.S. Quantum Computing Technical Performance (Metric IV.B.1)

Qubit Type	Physical Qubit Count	Readout Error	Qubit Coherence Time (microsec)	1-Qubit Gate Error Rate	1-Qubit Gate Time (ns)	2-Qubit Gate Error Rate	2-Qubit Gate Time (ns)
SC transmon (IBM)[a]	65	3.5×10^{-2}	122	3.8×10^{-4}	21	6.4×10^{-3}	199
SC transmon (Google)[b]	53	3.8×10^{-2}	25	1.2×10^{-3}	14	3.8×10^{-3}	28
Trapped ion (IonQ)[c]	20	1×10^{-4}	6×10^{8}	1.1×10^{-4}	2×10^{3}	6.7×10^{-3}	1×10^{4}
Trapped ion (Honeywell)[d]	10	2.5×10^{-3}	6×10^{8}	9×10^{-5}	NA	2.4×10^{-3}	NA
Quantum dot (Delft, Qutech, Intel)[e]	2+	NA	2800	1.0×10^{-3}	100	2×10^{-2}	NA

NOTES: Only qubits actually used in computations are included. NA = not available. All metrics presented here are current as of July 2021.

[a] Eric J. Zhang et al., "High-Fidelity Superconducting Quantum Processors via Laser-Annealing of Transmon Qubits," *ArXiv:2012.08475 [Quant-Ph]*, December 15, 2020; Petar Jurcevic et al., "Demonstration of Quantum Volume 64 on a Superconducting Quantum Computing System," *ArXiv:2008.08571 [Quant-Ph]*, September 4, 2020.

[b] Arute et al., 2019; B. Foxen, et al., "Demonstrating a Continuous Set of Two-Qubit Gates for Near-Term Quantum Algorithms," *Physical Review Letters*, Vol. 125, No. 12, September 15, 2020.

[c] K. Wright et al., "Benchmarking an 11-Qubit Quantum Computer," *Nature Communications*, Vol. 10, No. 1, November 29, 2019; Paul Smith-Goodson, "IonQ Releases a New 32-Qubit Trapped-Ion Quantum Computer with Massive Quantum Volume Claims," *Forbes*, October 7, 2020.

[d] Honeywell, "Honeywell Sets New Record for Quantum Computing Performance," March 2021; J. M. Pino et al., "Demonstration of the Trapped-Ion Quantum CCD Computer Architecture," *Nature*, Vol. 592, No. 7853, April 2021.

[e] Ruoyu Li et al., "A Crossbar Network for Silicon Quantum Dot Qubits | Science Advances," *Science Advances Magazine*, Vol. 4, No. 7, July 6, 2018.

capable the system potentially is. Key technical performance metrics for quantum circuit operations, including 1- and 2-qubit gate performance metrics, are also shown in the table. The gate times for 1- and 2-qubit gates refer to the amount of time that it takes to perform an elementary logic operation—the quantum equivalent of the AND, OR, XOR, or NOT gates from classical electrical engineering—on one or two qubits, respectively. The error rates indicate the probability that each elementary logic operation results in a mathematically incorrect output. Faster qubit gate operations and lower gate error rates are indications of superior system performance.

Qubit coherence time is another important metric. The longer qubits can remain coherent, the longer a quantum computer can be used for computation. Coherence times are much shorter for superconducting transmon (SCT) qubits than for trapped-ion qubits, but gate operation times are much faster in SCT systems. Therefore, both types of systems can perform a significant number of logic gate operations before the qubits decohere.[24]

Qubit state readout error is another important technical performance metric that corresponds to the probability that the qubit's final state is measured incorrectly at the end of the computation. Readout errors are smaller for trapped-ion systems than they are for SCT quantum computers. However, the number of qubits available for computation are significantly higher for SCT systems.

Recently, progress has been made in developing a quantum computer based on neutral cold atom qubits by researchers at Harvard, MIT, and the start-up ColdQuanta. These types of systems have the potential to execute long and complex calculations because of their long qubit coherence times. However, quantum circuit operations have yet to be publicly demonstrated in cold-atom systems.

The start-up PsiQuantum is attempting to develop a fault-tolerant and scalable photonic-qubit quantum computer capable of quantum error correction, using relatively commercially mature semiconductor fabrication equipment, but they had released very little public information about their progress as of our data collection cutoff. Microsoft has spent years researching the topological qubit paradigm but has not yet demonstrated any topological qubits. These companies are therefore not included in Table 3.25.

Until June 2021, the United States was the only country to have publicly documented performance for universal quantum-computer prototypes.[25] Since that time, China has published or posted claims of impressive results for three different types of quantum computers. We review these developments in the next chapter.

---

[24] These metrics cannot always be directly compared across different qubit types. For calculating how many logical operations can be performed before the qubits decohere, the relevant quantity is the *ratio* of the qubit coherence time to the gate operation times. So trapped-ion qubits can currently undergo many more logic gate operations than superconducting qubits can before decohering, even though they have much slower gate times.

[25] The Canadian company D-Wave has developed a sophisticated quantum annealer based on SCT qubit technology, but this is a special-purpose device that is not capable of universal quantum computation.

Quantum Communications

We have assessed the technical state of the art in three specific applications and enabling technologies for quantum communications. All three involve quantum entanglement, which is a critical requirement for the more-advanced applications of quantum communications that are not yet mature. The applications are discussed in more detail in Chapter One.

1. Efficient generation of high-quality entangled photon pairs. This is the critical first step in almost all quantum communications applications, as entangled photons are the most important physical communication channel for transmitting quantum information.
2. Long-distance entanglement-based quantum key distribution. Most currently deployed QKD systems do not utilize entanglement and are considered a low priority to U.S. policymakers. However, a "second-generation" form of QKD[26] that uses entanglement closes many of the security vulnerabilities of traditional QKD and could provide a stepping-stone toward directly networking quantum systems, which is considered the most promising long-term application of quantum communications.
3. Long-distance networking of quantum devices for quantum state sharing and transmission. This is the most sophisticated form of quantum communications. Unlike QKD where the final outputs are simply classical bits, in these systems the final transmitted output is a full quantum state. This application is a requirement for eventually networking together quantum computers or sensors.

Table 3.26 presents the technical state of the art in entangled photon generation across several metrics, and the demonstrating country. There are two major approaches to generated entangled photon pairs: the large majority of working systems today use a mature process known as *spontaneous parametric down-conversion* in which a single photon is pumped from a laser into a nonlinear crystal that converts it into a pair of entangled photons. A more-recent process in which an electric field is applied to a gallium arsenide quantum dot appears promising but has not yet been made scalable.

There are a variety of technical metrics for entanglement sources that are frequently reported. Briefly, the pair production efficiency refers to the probability that each attempt to produce an entangled pair on demand will succeed. The brightness refers to the total number of pairs that can be produced per second with a given amount of input power. The Hong-Ou-Mandel visibility and the fidelity measure the "quality" of each photon pair: the visibility measures the degree to which the photons are identical (a prerequisite for the interference effects that enable most optical quantum computing techniques, such as boson sampling), and the fidelity quantifies how close they are to being perfectly entangled. Which of these metrics is most important depends on the specific application.

---

[26] Known as *measurement-device-independent* QKD.

TABLE 3.26

Technical State of the Art in Entangled Photon Generation (Metric IV.B.1)

Generation Mechanism	Demonstrating Countries	Pair Production Efficiency	Brightness (Pairs/s/mW)	Indistinguish-ability (HOM Visibility)	Fidelity
Spontaneous parametric down-conversion	Germany[a]	43%	$3.5 * 10^6$	82%	96%
	Singapore[b]	N/R	$5.6 * 10^5$	98%	N/R
	United States[c]	20%	$2.7 * 10^9$	N/R	N/R
Gallium arsenide quantum dots	Germany[d]	37%	N/R	N/R	90%
	China[e]	65%	N/R	90%	88%

NOTES: HOM = Hong-Ou-Mandel, N/R = not reported. "Demonstrating country" refers to the location of the institutional affiliation of the authors (in the final row, the lead authors were affiliated with Chinese institutions, but there were European and U.S. coauthors). The four rightmost columns contain our technical metrics. For each metric, a higher number is better (holding all other metrics equal).

[a] Evan Meyer-Scott et al., "High-Performance Source of Spectrally Pure, Polarization Entangled Photon Pairs Based on Hybrid Integrated-Bulk Optics," *Optics Express*, Vol. 26, 2018.

[b] Alexander Lohrmann, Chithrabhanu Perumangatt, Aitor Villar, and Alexander Ling, "Broadband Pumped Polarization Entangled Photon-Pair Source in a Linear Beam Displacement Interferometer," *Applied Physics Letters*, Vol. 116, 2020.

[c] Zhaohui Ma et al., "Ultrabright Quantum Photon Sources on Chip," *Physical Review Letters*, Vol. 125, 2020.

[d] Y. Chen, M. Zopf, R. Keil, F. Ding, and O. G. Schmidt, "Highly-Efficient Extraction of Entangled Photons from Quantum Dots Using a Broadband Optical Antenna," *Nature Communications*, Vol. 9, 2018.

[e] J. Liu,et al., "A Solid-State Source of Strongly Entangled Photon Pairs with High Brightness and Indistinguishability," *Nature Nanotechnology*, Vol. 14, 2019.

Table 3.26 shows that there is no single leading country across all of these metrics; a different nation has produced some of the most advanced entanglement sources as measured by each metric. Notably, Germany is or near the global cutting edge for both classes of entanglement sources.[27]

The second quantum communications technology, long-distance entanglement-based QKD, does not appear to be a major R&D priority within the United States, and we were unable to identify any U.S. deployments near the world forefront.[28] We therefore defer discussion of this technology to Chapter Four, as China is making major efforts in this area.

Table 3.27 displays the technical state of the art in long-distance quantum networking, by which we mean the transmission of full quantum states (as opposed to classical bits, as with QKD), based on published academic literature. Currently, all quantum networking prototypes transmit qubits (in the form of photons) through fiber-optic cables, although this is not a physical requirement. The table indicates the longest length of cable that carried a successful transmission, and whether the cable was coiled up within a single laboratory or straight

---

[27] This conclusion agrees with input gathered from our conversations with industry, who consistently named Germany as a major source of high-quality and affordable photonics components.

[28] There are several U.S. start-ups working on QKD that have announced modest levels of VC funding, but we did not identify any documented technical performance, and most of them appear to be buying their equipment from abroad and focusing on service delivery.

TABLE 3.27

Technical State of the Art in Long-Distance Quantum Networking (Metric IV.B.1)

Endpoint Systems Entangled	Demonstrating Country	Entanglement Distribution Distance (Cable Configuration)	Entanglement Production Rate	Fidelity
Two photonic qubits	United States[a]	44 km (coiled)	1 pair/second	86%
One photonic qubit and one quantum memory	Austria[b]	50 km (coiled)	1 pair/second	86%
	China[c]	10 km (coiled)	2 pairs/second	78%
Two quantum memories	China[d]	50 km (coiled) / 22 km (straight)	Not reported	72% (straight fiber)

NOTES: "Demonstrating country" refers to the location of the institutional affiliation of the authors (in the first row, the lead authors were affiliated with the United States, but there were Canadian coauthors). The three rightmost columns contain our technical metrics. For each metric, a higher number is better (holding all other metrics equal).

[a] Raju Valivarthi et al., "Teleportation Systems Toward a Quantum Internet," *PRX Quantum 1*, 020317, December 4, 2020.

[b] V. Krutyanskiy et al., "Light-Matter Entanglement over 50 Km of Optical Fibre," *Nature Quantum Information*, Vol. 5, 2019, p. 72.

[c] W. Chang et al., "Long-Distance Entanglement Between a Multiplexed Quantum Memory and a Telecom Photon," *Physics Review*, Vol. 9, November 14, 2019, p. 041033.

[d] Yong Yu et al., "Entanglement of Two Quantum Memories via Fibers over Dozens of Kilometres," *Nature*, Vol. 578, 2020, pp. 240–245.

(giving a true physical end point separation). The entanglement production rate refers to the number of entangled pairs produced and transmitted per second, and the fidelity quantifies the quality of the entanglement between the two final end points (with 100 percent corresponding to a perfect match with the targeted maximally entangled two-qubit state).

This is a diverse and rapidly developing area, and different nations' capabilities are not yet standardized enough to enable a completely "apples-to-apples" comparison, so unfortunately, the rows in the table are not entirely directly comparable. Photonic qubits cannot be stored but must be used instantly for a one-shot application such as quantum teleportation. Quantum memories, on the other hand, are matter-based systems that are much more stable and allow for relatively long-term storage of a quantum state (currently up to almost one second). A quantum communications system with quantum memories at both ends is a requirement for quantum repeaters and other technologies that enable true long-distance networking of complex quantum systems such as computers or sensors.

However, the process of transferring a qubit state from a photon in motion to a stationary quantum memory (which is made of a completely different physical medium, such as trapped matter ions), known as "transduction," poses difficult technical challenges. Therefore, each quantum memory at either end point of the transmission greatly complicates the technical difficulty of the experiment and reduces the final fidelity between end points. We therefore assess that despite the fact that the last row in the table nominally has the lowest fidelity, it is actually the most technically impressive and application-relevant achievement by a significant degree.[29]

---

[29] It is also the only demonstration in the table that achieved a long physical separation between end points.

We did find an unpublished academic preprint by a U.S. group that demonstrated many of the necessary ingredients for entangling two far-separated quantum memories, but they did not claim to actually entangle stable stored states in the memories.[30]

Quantum Sensing

Sensors are conventionally described by performance-oriented metrics such as sensitivity, bandwidth, and dynamic range. Sensitivity measures a sensor's ability to detect whether a signal is present over a certain interval of time or to discriminate between unique frequencies in an incident signal. Bandwidth is associated with the minimum and maximum range of detectable frequencies (energy range) and their resolution. Dynamic range characterizes the strength of the sensor's response for a detected signal by contrasting the smallest and largest amplitudes of its response. Other measures that contribute to understanding the state of the start include how stable the sensor is with respect to noise and interference, its response time, operating requirements, and technological dependencies required for implementing the sensor.

There are a huge range of quantum sensors in development, and we were unable to survey them all. We chose to focus on gravimeters and magnetometers as case studies for comparisons, as both types of sensors represent a rapidly evolving and broadly active area of development for quantum sensing technologies. Other areas where advancements in quantum technology are moving the state of the art include timing (e.g., atomic clocks), bolometers for measuring radiant energy, and radio-frequency sensors. Although significant, the breadth of research and development activities are relatively limited across institutions, and a brief survey indicated that the United States appears to be at the forefront in each of these areas.

Developing broadly applicable metrics is also more challenging for quantum sensors than for quantum computers or communications because of the varied nature of the quantum improvement to sensors. Depending on the intended application, the relevant metrics might measure absolute sensitivity, or size, weight, power, and cost, or stability, or low-maintenance requirements—and only the first property is typically reported during the laboratory stage where most of these technologies currently remain. We therefore only captured fielding readiness at the qualitative level.

The United States is advancing the state of the art in several classes of gravimeters and is prioritizing deployability of these types of sensors. Many countries, including the United States and China, are developing new classes of highly sensitive magnetometers, such as nitrogen vacancy centers. Some technical metrics for leading examples of both types of sensors are summarized in Tables 3.28 and 3.29.

Existing, mature quantum-based sensors such as SQUID (superconducting quantum interference device) magnetometers still provide the best overall sensitivity but do so at the cost of versatility in sensing applications and limitations stemming from operating param-

---

[30] Dounan Du et al., "An Elementary 158 Km Long Quantum Network Connecting Room Temperature Quantum Memories," *ArXiv.org, Quantum Physics*, January 2021.

TABLE 3.28

Technical Achievement Metrics for Gravimeters (Metric IV.B.1)

Quantum System	Demonstrating Country	Sensitivity (nm/s² √Hz)	Drift Rate (µGal/month)	Operating Temperature	Fielding
Superconducting	USA[a]	3	0.5	4 K	Commercial
Cold atom	USA[b]	370	2,920	2 µK	Embeddable prototype
Cold atom	France[c]	500	1	2 µK	Embeddable prototype

NOTES: Holding all else equal, lower sensitivities and drift rates and higher operating temperatures indicate higher performance. The four rightmost columns contain our technical metrics.

[a] GWR Instruments, Inc. "iGRAV® Gravity Sensors," webpage, 2019.

[b] Xuejian Wu et al., "Gravity surveys using a mobile atom interferometer," *Science Advances*, Vol. 5, No. 9, 2019.

[c] Vincent Ménoret et al., "Gravity measurements below 10⁻⁹ *g* with a transportable absolute quantum gravimeter," *Nature Scientific Reports,* Vol. 8, 2018.

TABLE 3.29

Technical Achievement Metrics for Magnetometers (Metric IV.B.1)

Quantum System	Application	Demonstrating Country	Sensitivity µT/√Hz	Operating Temperature	Fielding
Atom vapor	Portable biomagnetometry	USA[a]	16×10^{-9}	Room temperature	Embeddable prototype
NV-center diamond	Biomagnetometry	Denmark, France, Germany[b]	10^{-4}	Room temperature	Laboratory
NV-center diamond	Magnetic and thermal imaging microscopy	China[c]	1.8	Room temperature	Laboratory
NV-center diamond	CMOS integrated sensing	USA[d]	32.1	Room temperature	Embeddable prototype

NOTES: CMOS = complementary metal-oxide-semiconductor, NV = nitrogen vacancy. Holding all else equal, lower sensitivities and drift rates and higher operating temperatures indicate higher performance. The three rightmost columns contain our technical metrics.

[a] M. E. Limes et al., "Portable Magnetometry for Detection of Biomagnetism in Ambient Environments," *Physical Review Applied*, Vol. 14, 2020.

[b] James L. Webb et al., "Optimization of a Diamond Nitrogen Vacancy Centre Magnetometer for Sensing of Biological Signals," *Frontiers of Physics*, 2020.

[c] Yulei Chen et al., "Simultaneous imaging of magnetic field and temperature using a wide-field quantum diamond microscope," *European Journal of Physics Quantum Technology*, Vol. 8, 2021.

[d] Donggyu Kim et al., "A CMOS-integrated quantum sensor based on nitrogen–vacancy centres," *Nature Electronics*, Vol. 2, 2019.

eters. In contrast, current R&D and commercial activity is focused on using new, alternative quantum particle systems that are demonstrating initial sensitivities with reasonable trade-offs and promising a much greater range of uses with fewer constraints and promising equal or greater sensitivities as their implementation matures. Room-temperature atomic vapor cells and nitrogen vacancy centers in diamond are new avenues for quantum technologies in magnetometry, while prototypes for new gravimeters focus on the use of cold atom systems.[31]

U.S. contributions are broad and emphasize component-level technology contributions using quantum systems with a very diverse range of future platforms based on distinct quantum systems (e.g., atomic vapor, electron gas, NV centers in diamond). Correspondingly, individual institutions may be exploring advances in sensor technology using different methods for the *same kind* of sensing devices, making it difficult to track advancement of state of the art in a specific area. Adding complexity to this picture, there is also a diverse *range* of potential applications for sensors—sensors having the same underlying technology, but tuned or prepared differently during the materials engineering process can support distinct sensing applications, arising from different sensitivities in the physics of the materials that have been fabricated.

C. Breadth of Technical Approaches Under Pursuit

Metric IV.C.1 concerns the number of distinct technologies for which the nation has deployed integrated prototypes with documented performance. In quantum computing, we have identified U.S. companies that have demonstrated and documented fully integrated prototypes for two qubit technologies capable of universal quantum computing: superconducting transmon qubits (Google, IBM, and Rigetti) and trapped-ion qubits (IonQ and Honeywell).[32] Other companies are working on other qubit technologies (e.g., PsiQuantum with photonic qubits, ColdQuanta with neutral-atom qubits, Intel with quantum-dot qubits, and Microsoft with topological qubits) but have not yet demonstrated integrated prototypes with clearly documented performance.

In quantum communications, there are several companies (some of which have received VC funding) attempting to deploy quantum key distribution commercially, but we were unable to determine whether they manufacture their own equipment and did not find documented performance metrics. We did not find evidence that the United States is advancing the state of the art in QKD. We did identify one company (Qunnect) attempting to deploy fully integrated "second-generation" quantum communications technology based on entanglement distribution, which has documented certain performance metrics as discussed above.

---

[31] However, atom interferometry with potential applications for gravimetry has also been demonstrated in a room-temperature vapor of rubidium atoms. G. W. Biedermann et al., "Atom Interferometry in a Warm Vapor," *Physical Review Letters*, Vol. 118, 2017.

[32] Other companies are working on these technologies as well but have not yet publicly demonstrated any prototypes with clearly documented performance metrics.

In quantum sensing, we identified 11 companies pursuing integrated hardware systems for six applications: gravimetry for PNT, radio-frequency sensing, magnetometry, gyroscopes, infrared sensing, and atomic clocks. We were unable to verify how many of them actually have commercially available products yet.

Metric IV.C.2 concerns the number of subdomains in which the nation is the technical world leader. This is inherently a somewhat subjective question, and different SMEs could draw different conclusions from the same data. With that caveat, we make the following judgments regarding quantum computing:

- As discussed in the next chapter, China has released (not yet peer-reviewed) preprints claiming impressive achievements with superconducting transmon qubits. If these claims are confirmed, then we judge that the United States and China are at rough parity in SCT technology. If they are not confirmed, then the United States is still the clear leader in publicly documented performance.
- The United States is the clear leader in trapped-ion quantum computers.
- China has documented the most advanced photonic quantum technology performance, but no country has demonstrated a system with true photonic qubits capable of universal computing. The U.S. firm PsiQuantum is working on such photonic qubits but has not released any performance data. However, PsiQuantum has raised hundreds of millions of dollars in VC, so it does have serious production capability.
- U.S. firms appears to be the closest to useful computers based on neutral-atom and quantum-dot qubits, but no nation has yet demonstrated a fully integrated prototype.
- U.S. firms appear to be making the largest investments in topological qubits, but no nation has yet demonstrated any qubits of this type (let alone integrated prototype computers).

In quantum communications, we did not identify any subareas in which the United States is a clear leader. Many nations, including the United States, China, and Germany, have demonstrated roughly comparable entanglement generation in the lab (although conversations with industry indicate that European entanglement sources are more affordable). The United States has not advanced the state of the art in QKD, and China appears to be the only nation to have entangled faraway quantum memories.

Performance in quantum sensing is very challenging to compare across nations, given the variety of different applications. From our noncomprehensive survey that focused on gravimeters and magnetometers, we assess that the United States is at the forefront of multiple quantum technologies for sensors. The United States is also a leader in practical deployment outside the lab—for example, it is prototyping embedded CMOS integrated sensors.

Metric IV.C.3 concerns the technologies for which companies (or other organizations) have officially released quantitative road maps (with timelines) for deployment. There is of course no guarantee that these timelines will be met, but a public commitment to certain timelines provides a useful source of accountability for firms and indicates confidence. We were able to find announced quantitative road maps only from quantum computing companies, mostly in terms of qubit counts.[33] All of the road maps that we found were released by American companies. Figure 3.16 summarizes these companies' achieved progress and announced goals for qubit counts.

FIGURE 3.16

Actual and Aspirational Qubit Counts Achieved by Various Companies (Metric IV.C.3)

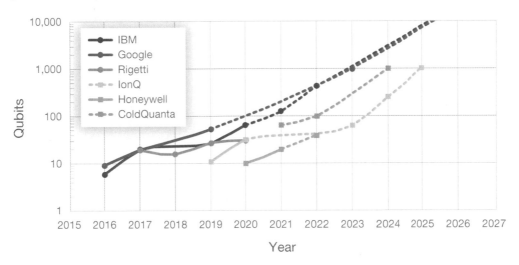

SOURCE: RAND analysis, based on Yu Chen, "Developing Technologies Towards an Error-Corrected Quantum Computer," speech delivered at IEEE Quantum Week 2020, held virtually, October 13, 2020; ColdQuanta, Inc., "Cold Atom Quantum Computing," video, YouTube, November 23, 2020; Jay Gambetta, "IBM's Roadmap for Scaling Quantum Technology," *IBM Research Blog*, September 15, 2020; Elizabeth Gibney, "Hello Quantum World! Google Published Landmark Quantum Supremacy Claim," *Nature*, October 23, 2019; Robert Hackett, "IBM Plans a Huge Leap in Superfast Quantum Computing by 2023," *Fortune*, September 15, 2020; Honeywell, "Get to Know Honeywell's Latest Quantum Computer System Model H1: Technical Details of Our Highest Performing System," webpage, undated; Abhinav Kandala et al., "Hardware-Efficient Variational Quantum Eigensolver for Small Molecules and Quantum Magnets," *Nature*, Vol. 549, No. 7671, September 14, 2017, p. 245; Patty Lee, "Unleashing the Power of Quantum for Everyone," speech delivered at the Future Compute Virtual Conference, held virtually, February 10, 2021; Ron Miller, "IBM Makes 20 Qubit Quantum Computing Machine Available as a Cloud Service," *Tech Crunch*, November 10, 2017; Moor Insights and Strategy and Paul Smith-Goodson, "IonQ Releases A New 32-Qubit Trapped-Ion Quantum Computer With Massive Quantum Volume Claims," *Forbes*, October 7, 2020; Mark Sullivan, "How IonQ Is Planning to Bring a Quantum Computer to the Masses," *Fast Company*, October 1, 2021; Rigetti, "What," webpage, undated; Brian Wang, "Google on Track to Make Quantum Computer Faster Than Classical Computers Within 7 Months," *Next Big Future*, June 23, 2017; and K. Wright et al., 2019.

NOTES: Solid curves denote achieved qubit counts (as of July 2021), and dashed curves represent announced goals. The logarithmic scale means that these companies anticipate an exponential increase in qubits over time (similar to Moore's law for classical transistors). This figure was truncated for clarity, but both Google and IBM have stated that they plan to reach one million qubits by 2030.

---

[33] There are a variety of possible explanations for this fact. It may indicate that the quantum computer firms are more confident in their technical prospects, or simply that qubit counts provide a relatively straightforward metric of progress, or that the firms feel a need to compete with each other for VC and public attention.

Summary of Findings

The United States has a very broad base of academic research, with over 1,500 institutions producing over 10,000 papers over the past decade (focusing on computing most, then communications, then sensing). Publishing in all three domains has seen steady growth. It produced more highly cited papers in computing and communications than any other country (although China produced more highly cited research in communications). Its research is highly international, with about half of all publications being international collaborations. A small but nonzero number of U.S. researchers collaborate with authors affiliated with strategic competitors' military universities.

The U.S. government is by far the largest research funder and is on track to spend $710 million on QIS R&D in FY 2021 from multiple agencies. Spending has grown at a steady rate of over 20 percent per year in recent years, largely driven by the National Quantum Initiative.

U.S. private industry in QIS is broad and diverse, with at least 182 firms—a mixture of large technology companies and recently founded start-ups—pursuing a wide variety of technology approaches and applications (focusing on computing most, then sensing, and relatively little on communications). VC is a very important source of financing for the start-ups, with $1.28 billion announced so far—the large majority of which has gone to just three firms. VC investment is heavily concentrated in quantum computing.

As of July 2021, the United States has documented the highest-performance prototypes in most technical approaches to quantum computing—except for the most mature approach (superconducting transmon qubits), in which China has claimed comparable performance in preprints still undergoing peer review. The United States is also a leader in the deployment of quantum sensing, but its R&D on communications remains primarily academic.

CHAPTER FOUR

China's Quantum Industrial Base

We have assessed that China is the country with the second most advanced industrial base in quantum technology after the United States; in most of the metrics that we applied across multiple countries, the United States and China took the top two spots (in either order). We therefore applied most of our metrics to the Chinese QIB as well, as a comparative case study and to demonstrate our framework's utility for comparing different nations. By design, this assessment was less comprehensive, but we believe that it paints an informative picture. As in Chapter Three, we used a uniform data collection cutoff of July 2021 for all of our metrics.

Assessment of Chinese QIS Research

A. Overall Research Activity

Table 4.1 depicts the number of publications on which an author affiliated with a Chinese organization for the three application domains.

TABLE 4.1
Total Chinese QIS Publications, 2011–2020 (Metric I.A.1)

Quantum Computing	Quantum Communications	Quantum Sensing
7,050	6,440	1,539

SOURCE: RAND analysis of Web of Science data.

B. Growth in Research Activity

Table 4.2 depicts the annual average rate of growth for Chinese publications for the three application domains.

TABLE 4.2
Compound Annual Growth Rate in Chinese Publications, 2011–2019 (Metric I.B.1)

Quantum Computing	Quantum Communications	Quantum Sensing
14.1%	8.9%	23.8%

SOURCE: RAND analysis of Web of Science data.

NOTE: Because the annual data for 2020 is not complete, we do not compute the annual growth rate between 2019 and 2020.

Figure 4.1 depicts China's annual publication output from 2011 to 2019 for all three application domains. As in the case of the United States, the graph depicts a positive growth trend in all three application domains, with China's totals doubling in all three application domains between 2011 and 2019.

FIGURE 4.1

Annual Growth in Chinese QIS Publications, 2011–2019

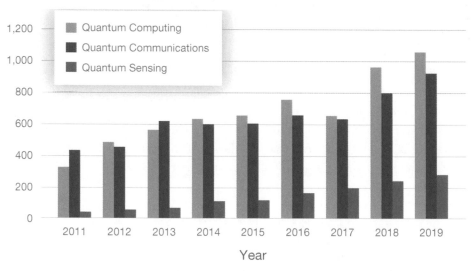

SOURCE: RAND analysis of Web of Science data.

C. Institutional Research Capacity

Table 4.3 depicts the number of Chinese research units to have produced a publication in each of the application domains over the 2011–2020 period of analysis.

TABLE 4.3

Number of Chinese Publishing Research Units, 2011–2020 (Metric I.C.1)

Quantum Computing	Quantum Communications	Quantum Sensing	Total
1,592	1,288	535	2,205

SOURCE: RAND analysis of Web of Science data.

Table 4.4 depicts the top-20 China-based research units by application domain.[1] As in the case of the United States, the list is comprised primarily of large research universities. Sixteen of the top 20 publishing organizations are universities. The Chinese Academy of Sciences,

---

[1] The 20 included research units are the top-20 publishing research units based on the sum of all three application domains. Research units are sorted by quantum computing.

the flagship national-level government research organization in China,[2] is the top global producer of scientific research within the three application domains considered here.

TABLE 4.4

Top 20 Chinese Research Units, 2011–2020

	Quantum Computing Publications	Quantum Communications Publications	Quantum Sensing Publications
Chinese Academy of Sciences, Beijing	556	348	103
Tsinghua University, Beijing	483	316	89
University of the Chinese Academy of Sciences, Beijing	266	99	107
University of Science and Technology of China, Hefei	225	307	69
Beijing University of Posts and Telecommunications, Beijing	217	558	12
Nanjing University, Jiangsu	213	77	34
Peking University, Beijing	199	122	24
Collaborative Innovation Center of Quantum Matter, Beijing	162	47	23
Shanghai Jiao Tong University, Shanghai	150	190	63
Beijing Computational Science Research Center, Beijing	131	53	30
Sun Yat Sen University, Guangdong	130	108	16
Zhejiang University, Zhejiang	125	63	56
Southeast University, Jiangsu	106	90	17
National University of Defense Technology, Hunan	98	120	29
Shanxi University, Shanxi	97	106	44
Shanxi University, Taiyuan	95	85	26
Dalian University of Technology, Dalian	95	95	10
Anhui University, Hefei	66	165	7
University of Science and Technology of China, Anhui	0	427	121
Chinese Academy of Sciences, Shanghai	0	122	101

SOURCE: RAND analysis of Web of Science data.

---

[2] The Chinese Academy of Sciences is comprised of over 100 research institutes, which tend to focus on a particular scientific or technical discipline.

Table 4.5 depicts the HHI for research units for China for the three application domains as well as for QIS. As in the case of the United States, research output in China is very uncon-centrated. As compared to the United States where MIT occupied the top slot for all three application domains, China has three distinct research units in first place in terms of highest concentration. The Chinese Academy of Sciences (Beijing) is listed as an affiliated research unit on 4.7 percent of Chinese quantum computing publications. Beijing University of Posts and Telecommunications has the highest share (5 percent) of quantum communications pub-lication slots. In quantum sensing, the University of Science and Technology of China, Anhui has the highest share; it is listed on 4.5 percent affiliation slots. As we will discuss below, we suspect that China's HHI has increased significantly in recent years, particularly for high-impact research, because starting around 2017, China has been concentrating much of its QIS research in the University of Science and Technology of China laboratory in Anhui. Although we decided to use the same assessment period across all of our metrics for methodological consistency, we believe that this increased concentration will be visible if these metrics are reassessed in the future.

TABLE 4.5

HHI for Chinese Research Units, 2011–2020 (Metric I.C.2)

Quantum Computing	Quantum Communications	Quantum Sensing	Total
0.0087	0.0098	0.013	0.0081

SOURCE: RAND analysis of Web of Science data.

D. Global Scientific Impact

Table 4.6 depicts China's highly cited publication output for the three application domains.

TABLE 4.6

Highly Cited Chinese Publications, 2011–2020 (Metric I.D.1)

Quantum Computing	Quantum Communications	Quantum Sensing
630	577	148

SOURCE: RAND analysis of Web of Science data.

Table 4.7 depicts the number of China-based research units to have produced at least one highly cited publication during the 2011–2020 period for the three application domains.

TABLE 4.7

Number of Chinese Research Units Producing Highly Cited Research, 2011–2020 (Metric I.D.2)

Quantum Computing	Quantum Communications	Quantum Sensing	Total
210	225	61	325

SOURCE: RAND analysis of Web of Science data.

E. Topical Alignment with Government Priorities

Table 4.8 depicts the percentage of publications on topics of low priority to the DoD, using the same inclusion criteria as in Chapter Three. Compared to the United States, a higher proportion of Chinese publications are in the low-DoD-priority subdomains. Over the 2010–2020 period of analysis, 34.9 percent of Chinese quantum communications publications were either QKD or quantum cryptography. Over same period, 41.4 percent of quantum sensing publications were on the topic of quantum imaging. This is more than twice the proportion of that of the United States. Figure 4.2 depicts the total publications for China across the three major application domains distinguishing for the subdomains (shaded in light blue and light orange) of low DoD priority.

TABLE 4.8

Percentage of Chinese Publications About Topics of Low Priority to DoD, 2011–2020 (Metric I.E.1)

Quantum Computing	Quantum Communications	Quantum Sensing
NA	34.9%	41.4%

SOURCE: RAND analysis of Web of Science data.

NOTE: NA, not applicable.

FIGURE 4.2

Total Chinese Publications by Application Domain with Low-DoD-Priority Subdomains

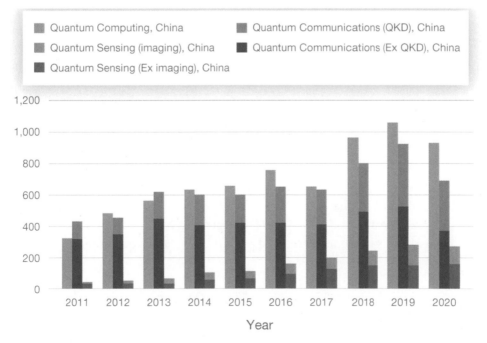

SOURCE: RAND analysis of Web of Science data.

F. Degree of Domestic and International Research Collaboration

Table 4.9 depicts the average number of collaborating domestic institutions per domestic institution (i.e., network degree) on Chinese publications for the three application domains.

TABLE 4.9

Average Number of Chinese-Collaborating Domestic Institutions, 2011–2020 (Metric I.F.1)

Quantum Computing	Quantum Communications	Quantum Sensing
9.77	10.84	6.52

SOURCE: RAND analysis of Web of Science data.

Table 4.10 depicts the percentage of Chinese publications that have an international coauthor. In all three application domains, the rate of international collaboration is lower among Chinese research units than their U.S. counterparts. For all three application domains, the United States is China's top collaborating country.

TABLE 4.10

Percentage of Chinese Publications Coauthored with Other Nations, 2011–2020 (Metric I.F.2)

Quantum Computing	Quantum Communications	Quantum Sensing
29.3%	18.7%	22.5%

SOURCE: RAND analysis of Web of Science data.

G. Risk of Technology Leakage

We did not assess these metrics for the Chinese QIB.

Assessment of Chinese Government Support

A. Overall Government R&D Investment

The total People's Republic of China (PRC) government investment in quantum R&D is challenging to estimate from public reporting, with different sources giving drastically different values. The most official estimates that we were able to find were from an academic article[3] from November 2019 coauthored by Jian-Wei Pan—the most prominent quantum technology researcher in the PRC, sometimes referred to as China's "father of quantum."[4] The reported

---

[3] Qiang Zhang, Feihu Xu, Li Li, Nai-Le Liu, and Jian-Wei Pan, "Quantum Information Research in China," *Quantum Science and Technology*, Vol. 4, No. 4, 2019.

[4] Martin Giles, "The Man Turning China into a Quantum Superpower," *MIT Technology Review*, December 19, 2018.

figures are summarized in Table 4.11, aggregated across five-year periods that correspond to the Chinese Five-Year Plans (FYPs), one of the PRC's main economic development initiatives. However, the Congressional Research Service estimated a much higher investment of $244 million for 2018 alone.[5]

TABLE 4.11

Reported Total QIS R&D Expenditures in China (Metric II.A.1)

Time Period	Estimated Annual QIS R&D Spending (USD)
1998–2006 (early stage)	$1.25 million
2006–2010 (11th FYP)	$30 million
2011–2015 (12th FYP)	$98 million
2016–2020 (13th FYP)	$84 million (as of November 2019)

SOURCE: Adapted from Zhang et al., 2019.

NOTE: Figures are not adjusted for purchasing-power parity (PPP). A correct adjustment for PPP is challenging to perform in a high-technology sector like quantum. The standard PPP adjustment for comparing nations' gross domestic products (GDPs) increased the purchasing power of Chinese spending by 71 percent above the exchange-rate conversion value during the 2011–2015 period, and by 63 percent during the 2016–2020 period. However, these GDP-wide conversions are not entirely applicable in a sector where the researchers are presumably paid much higher than the median Chinese worker, and we did not attempt to estimate a more precise adjustment.

Even higher levels of announced investment are associated with the main Chinese quantum research facility, the Hefei National Laboratory for Physical Sciences at the Microscale (HFNL), which is led by Pan and a part of the University of Science and Technology of China in Hefei, Anhui Province. Chinese-language news media reported $1.06 billion in laboratory funding in 2017,[6] and the *Anhui Business Daily* newspaper reported plans (though not confirmed funding) for $2.95 billion per year over the 2017–2022 period.[7]

These announced spending goals are in stark conflict with the government-wide spending estimates given in Table 4.11. The $1.06 billion start-up funding that Chinese news media announced in 2017 for Pan's quantum laboratory alone hugely exceeded Pan's own 2019 estimate for the PRC's *total* government spending over the same time period. Our team's China experts assess that these conflicting reports of funding levels are not unusual in China; the PRC government often announces ambitious (and often highly politicized) spending goals, and it is not uncommon for these goals to go unmet.

---

[5] Patricia Moloney Figliola, *Federal Quantum Information Science: An Overview*, Washington, D.C.: Congressional Research Service, IF10872, July 2, 2018.

[6] English-language media often report $10 billion in announced start-up funding for this laboratory (Arthur Herman, "At Last America Is Moving on Quantum," *Forbes*, August 20, 2018). We were unable to find any reference to this figure in Chinese-language media, and we believe that it stems from a common translation error involving the different structures of the English and Chinese counting systems.

[7] Elsa B. Kania and John K. Costello, *Quantum Hegemony? China's Ambitions and the Challenge to U.S. Innovation Leadership*, Washington, D.C., Center for a New American Security, September 12, 2018.

In summary, official reports of the PRC's government investment in quantum R&D in recent years have varied widely, from a low of $84 million per year (Pan's estimate) to a high of at least $3 billion per year (the *Anhui Business Daily*'s reported funding for Pan's laboratory). We are unable to assess from public information which figure is more accurate. By comparison, the U.S. government has spent $450–$710 million per year in recent years; we cannot determine whether the PRC total is higher or lower than this amount.

B. Growth in Government QIS R&D Investment

None of the data sources for Chinese QIS R&D investment that we found broke down the spending in enough detail for us to calculate annual growth rates. It is clear from Table 4.11 that investment grew enormously over the 2006–2015 decade. The lower levels reported since 2015 suggest that this growth may have leveled off or even reversed in recent years. On the other hand, the huge levels of announced funding for the HFNL suggest that investment is continuing to rapidly increase.

C. Stability of Government QIS R&D Investment

The PRC government budgeting system is much less publicly transparent than its U.S. equivalent. There is no precise Chinese equivalent to public legislation such as the U.S. National Quantum Initiative that formally authorizes multiyear appropriations for specific purposes. As such, there is no directly equivalent value for Metric II.C.1, the number and length of federal multiyear funding commitments.

However, the past two national Five-Year Plans have specifically mentioned quantum technology. The 13th FYP (covering the 2016–2020 period) mentioned quantum communications once, while the 14th FYP (covering 2021–2025) mentioned quantum technology seven times and described it as being of similar importance as other Chinese strategic priorities such as AI and advanced semiconductor manufacturing. These public documents indicate that quantum technology has been a sustained strategic priority of the PRC leadership that appears to be increasing in importance. As further evidence of senior leadership prioritization of quantum technology, PRC president Xi Jinping presided over a group study session on quantum science and technology in October 2020.[8]

---

[8] Huaxia, ed., "Xi Stresses Advancing Development of Quantum Science and Technology," *Xinhuanet. com*, October 17, 2020.

D. Breadth of Investment Sources

Table 4.12 displays the number of distinct funding sources to have funded at least 50 publications authored by Chinese organization–affiliated authors.

TABLE 4.12

Number of Distinct Significant Chinese QIS Research Funding Sources, 2011–2020 (Metric II.D.1)

Quantum Computing	Quantum Communications	Quantum Sensing	Total
24	27	7	32

SOURCE: RAND analysis of Web of Science data.

Table 4.13 depicts the HHI for Chinese funding sources. The overall HHI for funding sources is 0.273. This is substantially higher than that of the United States (0.141), indicating that funding of QIS in China is more concentrated than in the United States. As discussed in Chapter Three, the U.S. research funding distribution is as concentrated as a hypothetical distribution, with around seven equally important funders. By contrast, the Chinese funding distribution is as concentrated as a distribution, with fewer than four equal funders, indicating significantly more concentrated research funding. The HHI for funding sources in China is higher than that of the United States in all three application domains.

TABLE 4.13

HHI for Chinese Funding Sources, 2011–2020 (Metric II.D.2)

Quantum Computing	Quantum Communications	Quantum Sensing	Total
0.277	0.269	0.272	0.273

SOURCE: RAND analysis of Web of Science data.

The Chinese funding system is heavily dependent on the National Natural Science Foundation of China (NSFC). This organization is responsible for funding 50 percent of quantum computing publications, 50 percent of quantum communications publications, and 49 percent of quantum sensing publications in China. This is evident in Table 4.14, which depicts the number of publications funded by the top-20 QIS funding agencies in China. The table also illustrates the importance of province- and municipal-level funding in China. By contrast, state and local funding of scientific research is minimal in the United States (see Table 3.18, which contains no state-level funding agencies).

TABLE 4.14

Number of Chinese Publications Funded by Funder, 2011–2020

Funder	Quantum Computing Publications Funded	Quantum Communications Publications Funded	Quantum Sensing Publications Funded
NSFC	5,193	4,977	1,147
National Basic Research Program of China	1,115	761	199
Fundamental Research Funds for the Central Universities	722	788	141
National Key Research and Development Program of China	690	529	240
Chinese Academy of Sciences	604	447	156
China Postdoctoral Science Foundation	338	318	74
Ministry of Education	228	269	46
Specialized Research Fund for the Doctoral Program of Higher Education	198	301	42
Natural Science Foundation of Jiangsu Province	179	223	22
Program for New Century Excellent Talents in University	134	167	21
Beijing Natural Science Foundation	108	184	25
China Scholarship Council	148	122	28
Anhui Initiative in Quantum Information Technologies	130	113	47
Natural Science Foundation of Guangdong Province	145	124	2
National High Technology Research and Development Program of China (863 program)	61	126	68
Priority Academic Program Development of Jiangsu Higher Education Institutions	86	134	18
Natural Science Foundation of Anhui Province	63	135	19
Natural Science Foundation of Jiangxi Province	69	106	11
Ministry of Science and Technology	119	47	16
Natural Science Foundation of Shandong Province	60	81	22

SOURCE: RAND analysis of Web of Science data.

Assessment of Chinese Private Industry

A. Number and Distribution of QIB Firms

It is difficult to identify the number of Chinese firms that are meaningful participants in its QIB (Metric III.A.1). A 2020 Chinese business news report identified over 4,200 firms that *claim* to be working on quantum technology, of which 87 percent were established within the past five years.[9] However, we do not believe that this figure is an accurate reflection of the true Chinese QIB. The large majority of these companies appear to be "quantum" in name only, and Chinese SMEs have criticized the widespread abuse of the word in the commercial sector to capitalize on the current hype around the term. Jian-Wei Pan, for instance, has criticized companies claiming to sell "quantum skin care" products.[10] As a point of comparison, based on a search of Ph.D. dissertations related to quantum technology filed at Chinese universities, we estimate that only approximately 1,700 students have earned Ph.D.'s in quantum technology in China, so it is highly unlikely that a larger number of companies could be doing meaningful R&D in that field.[11]

As we were unable to fully characterize the companies in the Chinese QIB, we instead used a combination of Chinese-language news, English-language quantum industry publications and media reports, and patent searches to identify 13 Chinese firms that appear to be doing meaningful R&D in quantum technology.[12] These firms are listed in Table 4.15.

We were unable to find any public financial information on the large technology companies' quantum groups, but we were able to find information on the start-up companies from their websites and financial filings. The eight start-ups listed above form the (perhaps incomplete) data set for the rest of the metrics in this subsection. Figure 4.3 shows which application domains the start-ups work on.

Figures 4.4–4.6 show the distributions of start-up companies by reported employee count (Metric III.A.2), founding year (Metric III.A.3), and capital funding (Metric III.A.4), respectively. Given the limited data available and the fact that our search strategy did not capture

---

[9] Huanqiu Shibao (Global Times, Chinese edition), "Quantum Technology China Presses the 'Fast Forward Button' 4,200 Related Enterprises in the Quantum Field in My Country," October 22, 2020.

[10] Lin Zhijia, "Quantum Technology Commercialization Finds a Path, Capital Builds Momentum, but Technology Landing Pad Difficult," *Titanium Media APP*, in Chinese, March 16, 2021.

[11] We created this estimate based on data on Ph.D. dissertations available through the China National Knowledge Infrastructure (CNKI) database, which is roughly similar to ProQuest or LexisNexis for China. We searched CNKI for any Ph.D. dissertations that included the terms "quantum computing" (量子计算), "quantum communications" (量子通讯 or 量子通信), or "quantum sensing" (量子传感 or 量子精密测量) in either the title, abstract, or keywords, as provided by CNKI. The CNKI database does not include every Chinese university, but we believe that it provides a sufficiently large and representative sample to provide a useful rough estimate.

[12] This number is not directly comparable to the corresponding U.S. number reported in the previous chapter, which was assessed from a much more comprehensive data set.

TABLE 4.15

Major Chinese Firms Working on Quantum Technology R&D

Quantum-Focused Start-Ups
  Ciqtek (国仪量子)
  Kunfeng (昆峰量子)
  Origin Quantum (本源量子)
  Qasky (安徽问天量子科技)
  QuantumCtek (科大国盾量子技术股份有限公司 (国盾量子))
  QuDoor (国开启科量子技术 (北京) 有限公司), a.k.a. Qike Quantum (启科量子)
  Shenzhou Quantum Communication Technology (浙江神州量子通信技术有限公司)
  SpinQ (深圳量旋科技有限公司 (量旋科技))

Large Technology Companies with Quantum Research Groups
  Chinese Academy of Sciences-Alibaba Quantum Computing Laboratory (中国科学院–
  阿里巴巴量子计算实验室)
  Baidu Quantum Computing Laboratory (百度量子计算研究院)
  Huawei HiQ (Huawei Cloud) (量子计算软件云平台 (华为云))
  TenCent (腾讯量子实验室)
  ZTE (中兴通讯)

FIGURE 4.3

Distribution of Chinese Quantum Start-Ups by Application Domain (Metric III.A.1)

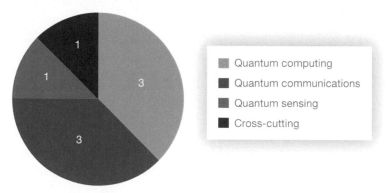

NOTE: One company (QuDoor Quantum) claims to be developing products in all three application domains and was categorized as "cross-cutting."

the basic component manufacturers, we did not attempt to characterize the distribution of firms across the production chain (Metric III.A.5).

Like the U.S. start-ups, the Chinese start-ups tend to be fairly small and young. The major difference between the U.S. and Chinese firms is that the Chinese companies have announced much, much lower capital funding. We were able to identify a total of only US$44 million in capital for the quantum start-ups, as compared to $1.28 billion for the U.S. start-ups.[13]

---

[13] We may have missed smaller quantum start-ups that would increase the Chinese total. But even if we only consider the top eight U.S. companies, their VC levels are vastly higher than the Chinese firms; in fact, the *average* VC level in the U.S. data set is higher than the *total* level in the Chinese data set.

FIGURE 4.4

Distribution of Chinese Quantum Start-Ups by Employee Count (Metric III.A.2)

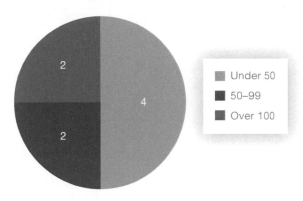

FIGURE 4.5

Distribution of Quantum-Focused Chinese Companies by Founding Year (Metric III.A.3)

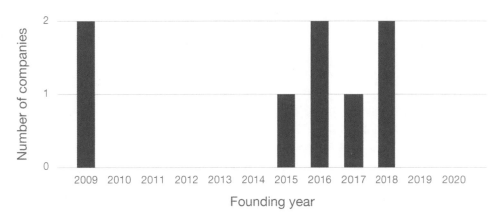

FIGURE 4.6

Distribution of Quantum-Focused QED-C Companies by Reported Capital Funding (Metric III.A.4)

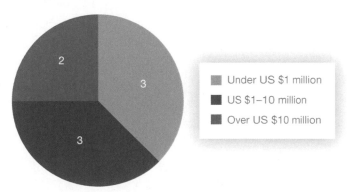

SOURCE: RAND analysis of Chinese financial filings.

Together with the fact that (unlike in the United States) the most advanced Chinese technical publications come from academic laboratories rather than private companies, we conclude that almost all quantum R&D in China appears to be controlled directly by the government.

We also found that three of the eight start-ups are headquartered in Hefei, and a fourth in another city in Anhui Province. These start-ups are presumably affiliated with the Hefei National Laboratory for Physical Sciences at the Microscale, further indicating the importance of this laboratory for the Chinese QIB.

B. Degree of Firm Specialization in Quantum Technology

As is the case in the United States, China's private-sector quantum industry contains both specialized start-ups and large technology companies (in apparently comparable numbers). Of the 13 firms that we identified as major players in quantum technology, eight are specialized start-ups; and five are large, diversified technology companies.

C. Foreign Dependencies

We did not attempt to assess foreign dependencies in the Chinese QIB's supply chains.

Assessment of Chinese Technical Metrics

A. Innovation Potential

This subsection presents the results of an analysis of data derived from the IFI CLAIMS Direct Platform global patent database regarding quantum technology patents filed within the PRC. We assessed the QIS patents filed in China using the same inclusion criteria as for those filed in the United States discussed in Chapter Three, a combination of topical keywords and technology classifications made by the issuing organization. Table 4.16 presents our top-level metrics for patents filed in China.

TABLE 4.16

Chinese Patenting Metrics in Quantum Technology Through 2019

Metric	Quantum Computing	Quantum Communications	Quantum Sensing
Total patent applications (Metric IV.A.1)	3,133	3,133	1,121
Number of unique patent assignees (Metric IV.A.2)	1,789	1,176	392
Annual growth in patent applications (Metric IV.A.3)	40%	38%	29%

SOURCE: RAND analysis of IFI CLAIMS Direct patent data.

NOTE: Annual growth in patent applications refers to the CAGR in applications filed over the 2010–2019 period.

Figure 4.7 shows the cumulative number of Chinese patents filed in each application domain from 2000 to 2020.

FIGURE 4.7

Cumulative Chinese Quantum Technology Patent Applications, 2000–2019

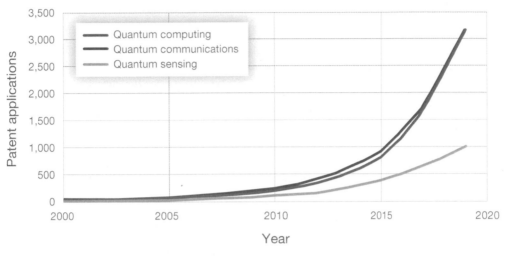

SOURCE: RAND analysis of IFI CLAIMS Direct patent data.

The Chinese patenting trends for quantum computing and communications are very similar. After a period of roughly linear increase, cumulative Chinese patent applications in both application domains began to increase exponentially around 2009—the start of the S-curve behavior typical of an emerging technology.[14] Although the number of quantum computing patent applications is less than that of the United States (3,133 compared to 4,845), the steeper rise of the Chinese S-curve suggests that China's total may soon equal or exceed that of the United States. By contrast, China has almost three times as many quantum communications patent applications as the United States. Although its S-curve emergence began about five years after that of the United States, its much steeper rise has enabled it to overtake the United States. Even though there are 1,176 unique assignees of these quantum communications patent applications, there are only 490 that filed more than one application.

We also see an S-curve emergence in quantum sensing beginning in 2009, rising from around 40 applications in that year to more than 650 in 2019. Although there are 392 unique assignees of these patent applications, there are only 137 that filed more than one application.

---

[14] Eusebi and Silberglitt, 2014.

B. Technical Achievement

Quantum Computing

We use the same technical metrics for quantum computers, with some minor extensions, that were used earlier to assess the capabilities of the U.S. QIB. These extensions were made to accommodate one specific technical approach being pursued by Chinese researchers that we did not include in the U.S. QIB assessment, as explained below.

The technical achievement metrics for Chinese government and academic institutions and companies are shown in Table 4.17. It appears the bulk of Chinese research on quantum computing is done at the Hefei National Laboratory for Physical Sciences at the Microscale. Researchers at this laboratory are pursuing a least three different technical approaches to quantum computing, as shown in the table:

- quantum processor powered by SCTs bonded directly to a cryogenic control microchip (second row in Table 4.17)
- quantum processor powered by SCTs with holes drilled into it that serve as conduits for 3D wire connections to a separate cryogenic logic chip (third row in Table 4.17)
- photonic quantum computer capable of bosonic sampling that has some programmability features (fourth row in Table 4.17).

All three of these quantum computing achievements are impressive, and all were made public in 2021. It should be noted that only the Zhong et al. paper has been published as

TABLE 4.17

Chinese Quantum Computing Technical Performance (Metric IV.B.1)

	Physical Qubit or Squeezed Photon Count	Qubit Readout Error	Qubit Coherence Time (microsec)	1-Qubit Gate Error	1-Qubit Gate Time (ns)	2-Qubit Gate Error	2-Qubit Gate Time (ns)
SCT[a]	66 (56)	4.5×10^{-2}	30.6	1.4×10^{-3}	25	5.9×10^{-3}	32
SCT[b]	62 (60)	5.8×10^{-2}	13.56	NYD	NYD	NYD	NYD
Photonic[c]	113	NA	NA	NA	NA	NA	NA

NOTES: NA = not available, NYD = not yet demonstrated, SCT = superconducting transmon. The table shows the number of physical qubits in each SCT quantum processor, with the parentheses indicating the number of qubits used to perform computations in each demonstration. The photonic computer demonstrated by Zhong et al. measured a maximum of 113 entangled photons, although these were not as computationally powerful as true qubits.

[a] Yulin Wu et al., "Strong Quantum Computational Advantage Using a Superconducting Quantum Processor," *ArXiv:2106.14734 [Quant-Ph]*, June 28, 2021. Another preprint posted September 2021 reported improved technical capabilities for this quantum computer, but we did not include this later preprint in our analysis as it was posted after our data collection cutoff. Qingling Zhu et al., "Quantum Computational Advantage via 60-Qubit 24-Cycle Random Circuit Sampling," *ArXiv:2109.03494 [Quant-Ph]*, September 9, 2021.

[b] Ming Gong et al., "Quantum Walks on a Programmable Two-Dimensional 62-Qubit Superconducting Processor," *Science,* Vol. 372, No. 6545, May 28, 2021.

[c] Han-Sen Zhong et al., "Phase-Programmable Gaussian Boson Sampling Using Stimulated Squeezed Light," *ArXiv:2106.15534 [Physics, Physics:Quant-Ph]*, July 5, 2021.

of July 2021.[15] The other two papers have been posted to the ArXiv preprint repository but have yet to complete peer review and be published. Only one of these devices (the Wu et al. quantum computer) has demonstrated the full array of capabilities for universal quantum computation. This device has a design that is similar to the Google quantum computer demonstrated in 2019 and has been used to perform the same type of computation that Google did in its 2019 landmark paper.[16]

Quantum Communications

As discussed in Chapter Three, we assessed the technical state of the art in three specific applications and enabling technologies for entanglement-based communications: efficient generation of high-quality entangled photon pairs, long-distance entanglement-based quantum key distribution, and long-distance networking of quantum devices for quantum state sharing and transmission.

For entangled photon pair generation, we refer the reader to Table 3.26 in Chapter Three and the discussion below it. We see that Chinese researchers have demonstrated gallium-arsenide quantum-dot entanglement sources of comparable quality to Germany's, but they have not demonstrated world-leading capabilities (according to our chosen metrics) for the more widely used spontaneous parametric down-conversion approach.

Table 4.18 presents the technical state of the art in very long-range entanglement-based QKD.[17] Three major delivery transmission channels for QKD have been demonstrated: fiber-

TABLE 4.18

Technical State of the Art in Long-Range Entanglement-Based QKD (Metric IV.B.1)

Transmission Channel	Demonstrating Countries	Entanglement Distribution Distance	Entanglement Production Rate	Fidelity
Fiber-optic cable	Austria[a]	96 km	257 pairs/second	Not measured
Satellite	China[b]	1200 km	1.1 pairs/second	87%

NOTES: "Demonstrating country" refers to the location of the institutional affiliation of the authors (in the first row, the lead authors were affiliated with Austrian institutions, but there were European and Chinese coauthors). For each metric, a higher number is better (holding all other metrics equal).

[a] Sören Wengerowsky et al., "Entanglement Distribution over a 96-Km-Long Submarine Optical Fiber," *PNAS,* Vol. 116, No. 14, April 2, 2019.

[b] J. Yin, et al., "Entanglement-Based Secure Quantum Cryptography over 1,120 Kilometres," *Nature,* Vol. 582, 2020.

---

[15] There is a published paper describing an earlier and less-capable prototype with a similar design: Han-Sen Zhong et al., "Quantum Computational Advantage Using Photons," *Science*, Vol. 370, No. 6523, December 18, 2020.

[16] Arute et al., 2019.

[17] Almost all existing QKD deployments do not use entanglement. It is much technically easier to deploy QKD without using entanglement, so the deployments without entanglement have higher nominal performance metrics. However, this mode has more security vulnerabilities and cannot be adapted toward quantum networks, so we do not track them here.

optic cable, atmospheric (i.e., open air directly between ground end points), and satellite. The longest demonstration over fiber-optic cable achieved a reasonably high transmission rate using a commercially deployed undersea fiber-optic cable connecting Sicily in Italy and Malta. We were unable to find any reports of long-distance open-air entanglement-based QKD since 2013, which we judged to be out of date. The only demonstration of satellite-based QKD was performed by the Chinese satellite *Mozi*, which (slowly) distributed entanglement over a vastly farther distance than any other demonstration, with reasonably good fidelity. Moreover, China has integrated its *Mozi* satellite into a sophisticated nationwide QKD network consisting of over 2,000 kilometers of fiber-optic cable.[18] Although the DoD and NSA have announced that they do not intend to deploy QKD themselves, it is worth noting that China's extensive deployment of QKD may help them develop technical expertise and production capacity for more impactful forms of quantum communications in the future. It could also help them develop capacity for deploying quantum technologies in other application domains, such as photonic quantum computing or single-photon quantum sensors, that require similar enabling technologies.

For long-distance quantum networking, we refer the reader to Table 3.27 in Chapter Three and the discussion below it. The main finding is that China is the only country to have demonstrated the long-distance entanglement of two quantum memories—an important step toward quantum networking.

Quantum Sensing

We did not attempt to do a deep dive into Chinese quantum sensing progress; we refer the reader to Tables 3.28 and 3.29 in Chapter Three and the discussion below. China does not appear to be advancing the state of the art in gravimetry, and its research in magnetometry appears to be focused on sample microscopy (e.g., for biomedical imaging) rather than for navigation or long-distance sensing.

C. Breadth of Technical Approaches Under Pursuit

Metric IV.C.1 concerns the number of distinct technologies for which the nation has demonstrated integrated prototypes with documented performance. As discussed above, in quantum computing, China has released a (not-yet-peer-reviewed) preprint claiming to demonstrate one integrated prototype using a universal qubit technology, superconducting transmon qubits. It has also demonstrated an integrated prototype for a boson sampling computing using photons as bosons, but this architecture is not capable of universal quantum computing and does not use any true qubits.

In quantum communications, China has deployed working prototypes (or better) of many key technologies: high-quality generation of entangled photon pairs, QKD over fiber-optic

---

[18] Yu-Ao Chen, Qiang Zhang, and Jian-Wei Pan, "An Integrated Space-to-Ground Quantum Communication Network over 4,600 Kilometres," *Nature*, Vol. 589, 2021.

cable, QKD (both with and without entanglement) and quantum teleportation via satellite, and the entanglement of two far-separated quantum memories.

Although we did not attempt a comprehensive survey, we did not identify any Chinese quantum sensor prototypes that are near practical fielding. Most Chinese research in quantum sensing appears to remain in the laboratory stage.

Metric IV.C.2 concerns the number of subdomains in which the nation is the technical world leader. For quantum computing, we refer the reader to the corresponding discussion in Chapter Three. To recap,

- If the claims in the Hefei laboratory's recently released preprint are verified, then China's deployment of SCT qubits is at rough parity with the United States'. If they are not verified, then the United States remains significantly ahead.
- China is the clear world leader in photonic boson sampling, although the practical applications of this paradigm are unclear.
- China does not lead in any other qubit technologies.

In quantum communications, we judge that China is at rough parity with several other nations (both in the United States and Europe, particularly Germany) in high-quality entanglement generation. It is the world leader in the large-scale deployment of QKD and the only nation to have demonstrated quantum communications by satellite. Moreover, as the only nation to have entangled quantum memories over long distances, it is the leader in quantum device networking—probably the most valuable application of quantum communications in the long term.

In quantum sensing, we did not identify any technologies in which China is a leader.[19]

We were unable to identify any publicly announced quantitative road maps with timelines to the useful deployment of any quantum technology from the Chinese government or industry (Metric IV.C.3).

Summary of Findings

Like the United States, China is a very active publisher of scientific research in all QIS application domains, with over 2,000 research units publishing over 14,000 publications over the past decade. The pace of research publication has increased steadily over the past decade. It produced fewer highly cited publications than the United States in quantum computing and sensing but more in quantum communications (and was in the top-two nations for all three domains). A significantly higher proportion of China's research focused on topics of low DoD

---

[19] This assessment agrees with an essay that Jian-Wei Pan wrote in a Chinese academic journal in which he assessed that "China started late in quantum sensing, and there is a certain gap with developed countries." Jian-Wei Pan, "Improve the Development of Our Nation's Quantum Technology," *Red Flag Manuscripts*, in Chinese, December 7, 2020.

priority than the United States' research. It has a significant level of international collaboration but less than the United States'.

Public reports of government investment in QIS R&D vary over orders of magnitude, from $84 million to almost $3 billion per year. Government funding is significantly more concentrated into a small number of agencies than the United States'. We are unable to determine from public sources whether China is spending more or less than the United States; however, senior Chinese leaders clearly regard quantum technology as a strategic priority.

Chinese QIS R&D appears to be dominated by government-funded laboratories, particularly the HFNL, based on recent claims of funding and technology achievement. Research at this laboratory is rapidly advancing the state of the art in several quantum technologies. By contrast, the private sector seems to play only a small role in cutting-end QIS R&D, with only 3 percent of the announced level of U.S. VC funding and few of the most significant technology demonstrations. The Chinese private sector is focusing much more on quantum communications than the U.S. private sector.

In terms of demonstrated technical achievement, China appears to be at rough parity with the United States in SCT qubits (the most mature quantum computing approach), and ahead in terms of the limited form of quantum computing known as boson sampling, but behind in other approaches. It is the world leader in multiple types of quantum communications, but is behind the United States in sensing.

CHAPTER FIVE

Findings and Recommendations

These findings and recommendations are not listed in order of importance.

Findings

1. The United States' overall scientific research output in quantum information science is broad, stable, and at or near the global forefront. It is too early to assess how successful its private sector will be at commercially fielding quantum products. Most quantum technologies are still at a very early stage of technology maturity. There are no clear applications expected for (at least) several years, and those eventual applications are still highly uncertain. We did not identify any critical gaps or immediate vulnerabilities in U.S. production capabilities, supply chains, commercial financing, or technical capabilities. However, with virtually no useful quantum products yet commercially available anywhere in the world, we do not have sufficient data to predict that no such gaps will emerge. The main areas of potential concern that we identified are China's demonstrated technical lead in certain quantum communications technologies and its very recent (still unverified) claims of having achieved rough technical parity in some forms of quantum computing.

2. The United States and China are the two world leaders in overall scientific research in each of the three quantum application domains. The United States leads in high-impact scientific publications in quantum computing and sensing. China leads in quantum communications. Moreover, China is continuing to widen its leadership in high-impact scientific publications in quantum communications. Much of this research is toward a specific technology (quantum key distribution) that is of low priority to U.S. policymakers, but some is toward entanglement-based technologies of potentially higher concern.

This difference reflects a broader difference in R&D focus between the two nations: the United States is focusing significantly more on computing than on communications or sensing, while China is focusing most on communications, with computing a close second. This pattern is consistently reflected across multiple lines of evidence: publication and patent quantity, the number and VC funding levels of private companies, demonstrated technical capability, and statements from national leadership.

3. The U.S. government is investing $710 million per year in quantum R&D (for FY 2021). This investment has grown rapidly (about 20 percent per year) in recent years.

Much of this funding growth is due to the National Quantum Initiative, which authorizes significant federal spending through 2023, but actual future appropriations are difficult to predict. We assess that significant federal R&D investment is still necessary for sustaining progress in quantum technology.

Chinese public reports of government quantum R&D funding vary widely, from $84 million to almost $3 billion per year, and we cannot assess from public reporting whether they are spending more or less than the United States.

4. U.S. quantum technology deployment is now driven by the private sector. U.S. private industry in quantum is broad and diverse, with a growing number of firms pursuing different technical approaches and no single leader. We have identified at least 182 firms that are involved in quantum technology to some extent, of which most of the quantum-specific firms have been founded since 2017. Most (although certainly not all) of these firms are working on quantum computing. There are multiple companies advancing the state of the art along completely different technology approaches, including both large technology companies such as Google, IBM, and Honeywell, and start-ups such as IonQ. There are two major sources of private financing: internal corporate R&D spending by large companies, whose levels are not public; and VC funding for start-ups, of which at least $1.28 billion has gone to at least 20 different firms. Federal government grants such as SBIR/STTR are a smaller (but still significant) source of funding for start-ups.

5. Chinese quantum R&D is concentrated in government-funded laboratories that have demonstrated rapid technical progress. The PRC has centralized most of its research efforts into one major laboratory in Hefei, which has recently announced (not-yet-peer-reviewed) quantum computing capabilities that match or surpass the U.S. state of the art. By contrast, we were able to identify only $44 million in announced Chinese private-sector capital (3 percent of the U.S. total), and we found few technological breakthroughs announced by private firms.

6. Some U.S. quantum technology firms are dependent on a small number of suppliers for high-quality components, most of which are located in Europe. This is particularly true of lasers and optical components (with Japan a key supplier of blue laser diodes). We did not identify any critical dependencies on strategic competitor nations, although some U.S. QIB companies are not certain of the ultimate origin of some of their components. Large high-quality sapphire wafers could potentially become a critical quantum material, and Russia is one of the few sources in the world for such wafers. There is no single supply chain for quantum technology; different technical approaches under investigation require completely different components, resulting in multiple supply chains with little overlap and distinct dependencies.

7. Imposing export controls on quantum computing and communications technology at this stage would slow scientific progress. These technologies are still at a low enough level of readiness that open scientific research is still a major driver of technology advancement, and many of the leading researchers are outside of the United States. QIS research is highly international, and export controls would impede this collaboration. No currently demonstrated quantum computing or communications technologies have immediate defense applications, nor is it yet possible to predict which technology approaches will eventually yield any.[1]

8. The United States has been leading in demonstrated technical capability in every major approach to quantum computing, although China has very recently announced progress that (if confirmed) brings it to rough parity in some approaches. China leads in demonstrated capability in quantum communications. Both the United States and Europe appear well positioned in quantum sensing. U.S. firms are demonstrating technical progress in every main technical approach to quantum computing, but China has recently announced comparable capabilities in superconducting-transmon qubits (one of the two leading approaches) and superior capabilities in photonic qubits, although not in trapped-ion qubits (the other leading approach). China has demonstrated technical leadership in quantum communications applications of both higher concern (networked quantum memories) and lower concern (QKD) to DoD policymakers.

9. There is a significant probability of quantum technology surprise. Technical capabilities in QIS are changing rapidly, and timelines to applications are highly uncertain. Technology surprise could play out in several different ways:

- There could be multiple changes in national technology leadership in the near- to mid-term future (probably, but not necessarily, between the United States and China).
- Multiple basic technical approaches are currently under development. The best approach could change, which would realign leadership within the industry (and possibly between nations).[2]
- There are no clear immediate-term applications for most current quantum technologies—particularly the current NISQ computers—so commercial demand remains limited.[3]

---

[1] Certain quantum sensing technologies are already export controlled, which we believe is appropriate.

[2] For example, the published academic literature indicates that superconducting-transmon qubits and trapped-ion qubits are currently the two most mature approaches for quantum computing. U.S. firms clearly lead in trapped-ion qubits, and either lead or are roughly comparable in SCT qubits (pending the verification of recent Chinese announcements). However, another U.S. firm, PsiQuantum, is working on a third approach (error-corrected photonic qubits) and has raised a huge amount of VC, so it could potentially achieve a technical breakthrough that moves it to the forefront. (It is worth noting that both China and Europe have demonstrated certain photonic qubit capabilities that surpass U.S. capabilities, so if photonic qubits become the new state of the art, other countries could gain a comparative advantage over the United States in quantum computing.)

There are also multiple fundamental quantum sensing approaches in development, but unlike quantum computers, different types of quantum sensors will probably always deliver fundamentally different capabilities, so it is unlikely that any single technical approach will win out.

[3] The only capability that existing quantum computer prototypes have demonstrated that could yield useful applications in the near future is the simulation of quantum physical systems (C. Neill et al., "Accurately Computing the Electronic Properties of a Quantum Ring," *Nature*, Vol. 594, 2021). However, these demonstrations are currently still of purely scientific interest.

But the discovery of a new application that can be delivered by current or near-term quantum technology (e.g., a clearly useful NISQ computer algorithm) could speed up development timelines by increasing commercial demand and focusing research effort.

- Some U.S. companies are working on developing challenging enabling technologies (such as topological or error-corrected qubits) that could drastically improve hardware capabilities. A breakthrough in any of these enabling technologies would drastically shorten the timelines to useful deployment. Conversely, hardware capabilities could also develop more slowly than currently anticipated.

Recommendations for Policymakers

1. Continue to provide a broad base of government R&D support across quantum technologies, complementing the most active areas of private investment. Given quantum's early stage of technical maturity and the high uncertainty in both the eventual applications and the best technical pathways, policymakers (including those at the major U.S. funding organizations listed in Table 3.18) should continue to fund a diversified portfolio of quantum technologies. Since academia and the private sector are both currently advancing the state of the art, a healthy ecosystem will require both public and private investments over the short and medium term.

When planning their overall investment portfolio, policymakers should be aware of commercial activity and should ensure that all strategic government priorities are being supported by the private and/or public sectors. For example, most U.S. private-sector investments currently go to quantum computing; policymakers should ensure that government funding also continues to sustain sufficient expertise and capital in quantum communications and sensing as well, as the commercial support for these domains is currently less robust. Even if the U.S. government does not anticipate ever deploying QKD, quantum communications technology will likely be an important enabler for larger quantum computers and networked quantum sensors, so we recommend that the U.S. government continue to support a research base in those technologies (e.g., through programs such as the DOE's recent $25 million quantum internet test bed program).

2. Monitor, and if possible, help protect, the quantum technology programs of key U.S. quantum technology firms. Many of the major quantum technology advancements, particularly in computing and sensing, are currently being made by a few key private-sector companies. The U.S. government should monitor these commercial programs for important advances and setbacks. If important advances are made, the government may need to help prevent or minimize technology theft and leakage to strategic competitors. The U.S. government should share foreign threat information with commercial companies (as appropriate) to help them increase their defenses or orient them to address specific technology leakage vectors.

3. Monitor the financial health and ownership of quantum start-up companies. Given the early stage of quantum technology, there are not yet any clear near-term commercial applications. Therefore, commercial start-ups in quantum technology will not have a clear revenue stream for the foreseeable future. The larger start-ups appear to be primarily supported by VC—which could dry up quickly if no applications become technically feasible—and most of the VC funding is going to only a few companies. To the extent possible, policymakers should attempt to monitor for financial risks to key quantum companies (such as a rapid decrease in VC or revenue) that could greatly disrupt the nascent U.S. quantum industry.

The U.S. government should also monitor the acquisition of small quantum technology companies by foreign nations. A few European companies have gained leadership in advanced laser technologies by acquiring small U.S. firms. Although not an immediate concern, this area should be monitored.

4. Monitor the international flows of key elements of the industrial base, such as critical components and materials, skilled workers, and final quantum technology products. Many of the highest-quality components for quantum technologies come from outside of the United States. We did not identify any critical supply-chain dependencies on strategic competitor nations, but this could change in the future. For now, policymakers should attempt to monitor the multiple distinct supply chains for the different technical approaches under pursuit. Once the technologies mature and the eventual applications become clearer, policymakers may need to narrow their scope to prioritize access to the specific quantum technologies most likely to support the government's strategic goals.

Quantum information science is an area of highly international technical collaboration, and much of the top technical talent comes from outside the United States. This international collaboration is very important for supporting the free flow of scientific information and for advancing the technical state of the art. But our assessment found some international research collaboration with Chinese institutions—particularly military-affiliated universities—that we believe poses a risk of technology leakage. As the two global leaders in all three QIS application domains, collaboration between the United States and China is to be expected, and it can yield positive political as well as scientific outcomes,[4] so we do not recommend restriction of this collaboration. However, the risk of scientific and technological leakage should be monitored.

5. Do not impose export controls on quantum computers or quantum communications systems at this time. Export controls would prematurely limit the exchange of scientific ideas, slowing down technological progress.[5] Having a broad base of experts (including outside the United States) experimenting with early-stage prototypes could speed up the discovery of useful defense-related applications. Moreover, export controls could threaten the

---

[4] Olga Krasnyak, "Science Diplomacy and Soviet-American Academic and Technical Exchanges," *The Hague Journal of Diplomacy*, Vol. 15, No. 3, 2020.

[5] This recommendation only applies to export controls placed on technology items. It does not apply to restrictions that the U.S. government might impose on exports to specific parties. For example, the Commerce Department added HFNL and QuantumCTek to its Entity List in November 2021, thereby restricting U.S. businesses from exporting to those organizations; our recommendation does not apply to these types of export controls. U.S. Department of Commerce, Bureau of Industry and Security, Final Rule 86 FR 67317.

financial health of small U.S. start-ups that are advancing the state of the art in quantum technology, as it is not clear that there will be enough domestic demand to support them.

Given the current uncertainty in eventual applications, we believe that at this stage, it would be impossible to craft export controls that apply to only the specific quantum computing and communications technologies that threaten U.S. national security. But once the technology advances closer to useful applications such as code-breaking, policymakers at the Commerce and State Departments (in consultation with other federal agencies) should reassess the need for export controls.

However, the U.S. government should ask domestic manufacturers to privately report their sales of quantum computing and communications equipment overseas to help the government monitor their deployment status per the previous recommendation.

6. Periodically reassess the rapidly changing quantum industrial base. The global QIB is growing quickly, but the true "killer apps" are still many years away, so the QIB is far from mature. One of the most challenging aspects of assessing the industrial base in a sector as young as quantum is determining a baseline for comparison: it is difficult to assess its strength at a single point in time without being able to track trends over time or learn from a successful deployment of quantum technology. Policymakers should regularly reassess the state of the QIB (perhaps using metrics such as those developed in this report) in order to identify positive or negative trends that may require policy responses. In addition to tracking the industrial bases of the United States and allied countries, policymakers should also track non-allied countries' progress toward final products and strategic applications. This is particularly true today in the domain of quantum communications, where China (a strategic competitor nation) has already demonstrated some capabilities that exceed the United States'.

Methodology Details

Research Metrics Methodology

The publication data used here are from the Core Collection of the Web of Science scientific publication database. The Web of Science is a database comprised of over 90 million records from 21,000 peer-reviewed journals. Journals included in the Web of Science database are curated based on quality and influence.[1] The low-quality, often pay-to-publish journals that have proliferated in recent years are not indexed in the Web of Science and thus do not influence the results of our analysis.[2]

Publication Search Strategy

As discussed in Chapter One, we have identified three major application domains in QIS: quantum computing, quantum communications, and quantum sensing. The objective of the publication search strategies used here is to build a query that yields the largest publications set possible that still matches the target application domain. The ideal sample for a given application domain would contain all of the scientific publications within the target domain and no publications that fall outside of the domain. In an attempt to arrive as close as possible to this ideal, we use a search strategy based on application-specific key terms that have been curated by SMEs in QIS.

Our search strategy for quantum computing is depicted in Figure A.1. The search strategies for the other two major application domains—quantum communications and quantum sensing—follow the same general process. In step 1 of the search strategy, researchers build a large set of candidate key terms. This list is generated by extracting all author-provided keywords from a set of known quantum computing publications.[3] The set of known quantum

---

[1] Web of Science Group, "Web of Science Core Collection," *Clarivate.com*, webpage, undated.

[2] Beall (2016) documents the proliferation of low-quality, pay-to-publish journals. Jeffrey Beall, "Essential Information About Predatory Publishers and Journals," *International Higher Education*, Vol. 86, 2016.

[3] Author-provided key terms are the terms included by a publications author to classify the publication according to the subject matter.

FIGURE A.1
Search Design Strategy

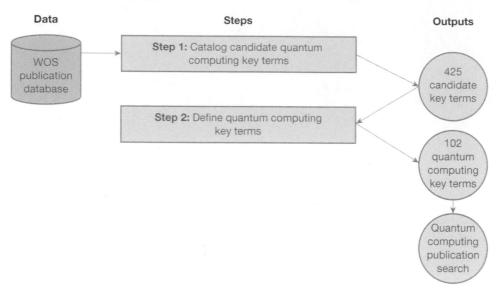

computing publications (i.e., the "seed" data set) is generated by searching the Web of Science database for a small set of terms that are clearly within the target application domain. For example, the initial terms used to define the quantum computing "seed" data set was simply ""Quantum Comput*"".

The set of author-provided keywords that are extracted from the initial "seed" publication data set are then tabulated based on the frequency with which they appear in the "seed" publication data set. All author-provided keywords that appear in more than five publications in the "seed" data set are provided to SMEs for assessment.

During step 2 of the search strategy, this set of candidate terms is independently evaluated by two SMEs according to a "minimize false positives" criterion. That is, the SMEs are asked to code as "yes" terms that are likely to be used in quantum computing publications but *not* used in other application domains. This criterion seeks to eliminate out-of-scope terms (e.g., "quantum key distribution" is outside of scope for quantum computing) and eliminate overly broad terms such as "quantum." Both SMEs coded the candidate terms independently. Terms that were coded as "yes" by both SMEs were included in the final search. For cases in which one SME coded a term "yes" and the other coded that term "no," the SMEs reached a consensus via discussion about how to code the term in question. For the quantum computing search strategy depicted in Figure A.1, of the 425 candidate terms, 102 were selected by the SMEs as appropriate terms for the quantum computing publication search. All of these terms were then used in the final quantum computing publication search.

This process was executed three times, once for each of the major application domains. In total, our QIS data set consisted of over 46,000 unique publications. Table A.1 depicts the final search terms for each of the three major application domains. These terms were used to construct a Boolean search string queried on the Web of Science database. When appropriate, wildcard (*) operators were added to the terms to account for alternative word endings (e.g., a search for "quantum cryptogr*" would yield results for both "quantum cryptography" and "quantum cryptographic"). The use of quotation marks notes the requirement for an exact match.

Besides the three major application domains, Table A.1 depicts the terms used to define certain subdomains of interest. Within the quantum communications application domain, we were interested in distinguishing QKD and quantum cryptography publications from the rest of the quantum communications domain. The search named "Quantum Communications (QKD and quantum cryptography)" refers to the search used to define the QKD and quantum cryptography subset. The search named "Quantum Communications (ex QKD and quantum cryptography)" was used to define the rest of the quantum communications appli-

TABLE A.1

Search Terms for Major Application Domains

Application Domain	Terms Included in Search
Quantum Computing	"adiabatic quantum comput*", "amplitude amplification", "analog quantum simulation*", "blind quantum comput*", "boson sampling", "bqp", "bqp-complete", "charge qubit*", "circuit quantum electrodynamics", "cluster state*", "d-wave", "delegated quantum comput*", "deutsch-jozsa algorithm*", "distributed quantum comput*", "duality quantum comput*", "durr-hoyer algorithm*", "fault-tolerant quantum comput*", "flux qubit*", "geometric quantum comput*", "grover algorithm*", "grover's algorithm*", "grover's quantum search algorithm*", "hadamard gate*", "hhl algorithm*", "holonomic quantum comput*", "linear optical quantum comput*", "logical qubit*", "measurement-based quantum comput*", "nisq", "nmr quantum comput*", "noisy intermediate scale quantum", "one-way quantum comput*", "optical comput*", "qaoa", "quantum advantage", "quantum algorithm*", "quantum annealing", "quantum approximate optimization algorithm*", "quantum automata", "quantum cellular automata", "quantum circuit*", "quantum compilation", "quantum compiler*", "quantum complexity", "quantum complexity theory", "quantum comput*", "quantum computation and information", "quantum computation architectures and implementation*", "quantum computational complexity", "quantum computational logic*", "quantum computer simulation*", "quantum computing simulation*", "quantum cost*", "quantum counting algorithm*", "quantum decryption", "quantum error correction", "quantum evolutionary algorithm*", "quantum finite automata", "quantum fourier transform*", "quantum game*", "quantum gate*", "quantum genetic algorithm*", "quantum image proces*", "quantum information proces*", "quantum knot*", "quantum lattice gas automata", "quantum logic gate*", "quantum logic synthesis", "quantum logic*", "quantum machine learning", "quantum neural network*", "quantum neuron*", "quantum optimization", "quantum parallelism", "quantum phase estimation algorithm*", "quantum private comparison", "quantum programming", "quantum programming languages", "quantum query algorithm*", "quantum query complexity", "quantum recommendation", "quantum register*", "quantum search algorithm*", "quantum search*", "quantum simulation*", "quantum software*", "quantum speedup", "quantum supremacy", "quantum turing machine*", "quantum verification", "quantum volume*", "quantum walk*", "shor's algorithm", "superconducting quantum comput*", "superconducting qubit*", "surface code", "topological quantum comput*", "topological qubit*", "universal quantum comput*", "variational quantum eigensolver", "variational quantum unsampling", "vqe"

Table A.1—Continued

Application Domain	Terms Included in Search
Quantum Communications (ex QKD and quantum cryptography)	"bell inequalities", "bell inequality", "bell state*", "bell state measurement", "bell states", "controlled quantum communication*", "entanglement concentration*", "entanglement distillation*", "entanglement distribution", "entanglement swap*", "epr pair*", "free-space quantum communication*", "heralded single photon source*", "heralded single-photon source*", "long-distance quantum communication*", "qber", "quantum bit commitment", "quantum bit error rate*", "quantum channel*", "quantum communication*", "quantum communication channel*", "quantum communication complexity", "quantum communication network*", "quantum communications", "quantum dense coding*", "quantum dialogue", "quantum direct communication*", "quantum discord", "quantum internet", "quantum key distribution*", "quantum network*", "quantum networks", "quantum private quer*", "quantum repeater*", "quantum repeaters", "quantum router*", "quantum sealed-bid auction*", "quantum shannon theor*", "quantum state sharing", "quantum teleportation", "remote state preparation*", "superdense coding*", "the bell state measurement*"
Quantum Communications (QKD and quantum cryptography)[a]	"quantum cryptogr*", "semi-quantum cryptogr*", "quantum secret sharing", "controlled quantum secure direct communication*", "quantum secure direct communication*", "deterministic secret quantum communication*", "deterministic secure quantum communication*", "quantum signature*", "quantum blind signature*", "quantum private comparison*", "quantum encryp*", "quantum authentication", "quantum identity authentication*", "secure quantum communication*", "arbitrated quantum signature*", "quantum secure communication*", "qsdc", "quantum communication security", "y-00 protocol*", "quantum steganogra*", "continuous variable quantum key distribution*", "continuous-variable quantum key distribution*", "quantum key distribution*", "measurement-device-independent quantum key distribution*", "qkd", "qkd network*", "b92", "b92 protocol*", "bb84", "bb84 protocol*", "decoy state*", "quantum key agreement", "measurement device independent", "measurement-device-independent", "semi-quantum key distribution*", "decoy state protocol*", "decoy states*", "quantum one-time pad*", "quantum key distribution network*", "quantum key distribution protocol*", "photon number splitting attack*"
Quantum Sensing (ex imaging)	"quantum sensing", "quantum sensor*", "quantum metrology", "atom interferometry", "n00n state*", "atomic sensor*", "quantum gyroscope*", "quantum accelerometer*", "quantum ins", "quantum imu", "quantum magnetometer*", "quantum rf receiver*", "cold-atom interferometer*", "cold-atom gas interferometer*", "heisenberg limit*", "standard quantum limit*", "quantum inertial sens*", "quantum gravimeter*", "quantum electrometer*", "quantum radio*", "quantum receiver*", "rydberg atom sensor*", "vapor-cell sensor*", "defect-based sensor*", "scanning quantum dot microsco*", "qubit detector*", "quantum detector*", "quantum detector tomography", "quantum tomography", "quantum state tomography", "microwave bolometer*", "microwave bolometer*"
Quantum Sensing (imaging)	"quantum illumination", "ghost imaging", "quantum dot imaging", "quantum imaging", "quantum radar*"

[a] We have combined QKD terms and quantum cryptography terms for our analysis.

cation domain. The union of these searches constitutes the full quantum communications application domain search. We were also interested in distinguishing quantum imaging from the rest of the quantum sensing application domain. The search named "Quantum Sensing (imaging)" refers to the search used to define the quantum imaging subdomain. The search named "Quantum Sensing (ex imaging)" was used to define the rest of the quantum sensing domain. Again, the union of these two searches constitutes the full quantum sensing application domain search.

In the publication analysis contained in the report, the nationality of a publication is determined by the "country" field within the Web of Science database. The "country" field is populated based on the affiliation address that an author lists during the publication process. For instance, a publication for which an author lists her address as "College of Computing, Georgia Institute of Technology, North Ave. NW, Atlanta, GA 30332, USA" will be assigned to the United States.

Additional Methodological Details on Research Metrics

I.A Overall Research Activity

Overall research activity is calculated as the number of publications on which an organization located within a given country is listed as an author's affiliation. Because many scientific publications are coauthored by individuals from distinct organizations, a portion of articles will be counted more than once. For example, a publication with two authors, one based in a Chinese university and one based in a U.S. university, will be counted in both countries' total publications tally.[4]

I.B Growth in Research Activity

Growth in research is calculated as the year-on-year percent change in total publications (Metric A.1). For example, if a country has 80 publications in year 1 and 100 publications in year 2, the year 2 growth in publications is 25 percent ((100 − 80)/80)). When the CAGR is reported, we use the formula

$$\text{CAGR} = ((\text{final value})/(\text{initial value}))^{1/T} - 1,$$

where T denotes the number of years between the initial and final value.

I.C Institutional Research Capacity

Metric I.C.1 is calculated as the number of unique research units that have been listed as an author's affiliation on a publication in a focal research area. We define a research unit as a group of geographically colocated researchers that share an organizational affiliation. Dis-

---

[4] An alternative calculation method would be to take fractional counts. In the example of a two-author Chinese-U.S. collaboration, a fractional count method would assign each country 0.5 for the publication. However, we have selected our method because it enables a more-intuitive interpretation.

tinguishing between geographically distinct units of an organization seeks to account for the large number of semiautonomous research organizations that are often organizationally located within a given "affiliation." Classifying each of these units under the parent affiliation would result in reduced fidelity to the organizational ecosystem that this measure seeks to partially measure. By way of example, the method used here distinguishes between publications that list "Google, Los Angeles, CA, USA" and "Google, Mountain View, CA, USA" in the author affiliation field.

Metric I.C.2 is calculated as the HHI for research units. The HHI is calculated as the sum of the squared publication shares of the research units within the scientific publication system. More formally,

$$HHI = \sum_1^N s_i^2,$$ where s_i is publications share of the ith research unit.

Publication share refers to the ratio of the number of publications on which a given research unit is listed as an affiliation and the total number of domestic affiliation slots for a given application domain. For example, for the quantum sensing application domain, the research unit "MIT, Cambridge, MA, USA" is listed as an affiliation 149 times, and there are 1,961 U.S. quantum sensing affiliation slots. Therefore, the research unit "MIT, Cambridge, MA, USA" has a publication share of 7.6 percent (149/1,961 = 0.076) for quantum sensing. This proportion is calculated for all research units.

The sum of the squared publication shares constitutes the HHI for the application domain in question. The higher a country's HHI, the more concentrated is its research capacity. If all of a country's publications from a given research area came from a single research unit, the country's HHI would be 1. As a country's research output is spread out across a larger number of research units, its HHI approaches 0.

When the N firms in a market all have equal share, the HHI equals $1/N$. Turning this observation around, we find that the reciprocal of the HHI for a market gives the number of firms in a hypothetical market that is as concentrated as the market in question overall but in which every firm has equal share. The reciprocal of the HHI is therefore sometimes referred to as the "effective number of firms in the industry." Very roughly, we can think of the reciprocal of the HHI as quantifying the equivalent number of "major players" in the market for the purpose of quantifying market concentration. This reciprocal quantity does not correspond to actual specific firms, but it gives a heuristic interpretation of the meaning of the HHI.

I.D Global Scientific Impact

Metric I.D.1 is calculated as the number of publications written by authors that are affiliated with an organization located within the focal country that fall into the top decile in terms of citations received during a given year. To account for the fact that citations accumulate, partially, as a function of time, the decile cutoff point is calculated on an annual basis.

Metric I.D.2 is calculated as the number of unique research units to have been listed as an author's affiliation on a top-decile publication in a focal research area. To account for the fact that citations accumulate, partially, as a function of time, the decile cutoff point is calculated on an annual basis. Figure A.2 depicts the citations frequency distribution for quantum computing publication published in 2017. The modal number of citations received by quantum computing publications published in 2017 is 0 (491 publications or roughly 19 percent of all publications have received 0 citations by the date of the query).

The measurement of highly cited publications seeks to address the fact that there is heterogeneity in publication impact. As scientists write up the results of their research, the norms of scientific documentation require them to cite prior articles on which they have built.[5] Thus, the number of times that a given publication has been cited is an indicator of the impact that a given publication has had on subsequent scientific discovery.[6] This logic is supported by empirical research finding a correlation between citations and peer ratings of research.[7] Although citations are by no means a perfect metric of research impact, they are used here to complement publication counts and to account for variability in research impact.

The measurement of highly cited publications also seeks to account for systematic country-level variation in the impact of published scientific journal articles. Recent research has found that China's provision of direct financial incentives to researchers to publish has

FIGURE A.2

Frequency Distribution of Citations for Quantum Computing Publication in 2017

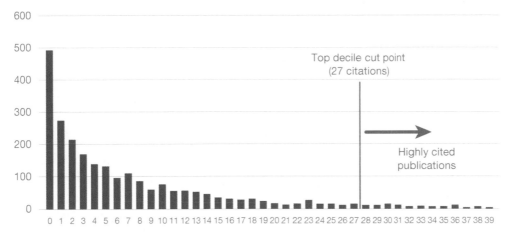

NOTE: The full right-hand tail of the distribution is not shown due to space constraints. The maximum number of citations received by any quantum computing publication published in 2017 was 1,256.

[5] R. K. Merton, "Foreword," in E. Garfield, ed., *Citation Indexing—Its Theory and Application in Science, Technology, and Humanities*, New York: John Wiley, 1979.

[6] Aksnes, Langfeldt, and Wouters, 2019.

[7] E. J. Rinia et al., "Comparative Analysis of a Set of Bibliometric Indicators and Central Peer Review Criteria: Evaluation of Condensed Matter Physics in the Netherlands," *Research Policy*, Vol. 27, 1998.

resulted in a glut in low-quality publishing in that country.[8] Systematic country-level variation in publication impact suggests that simple publication counts will suffer from low cross-country metric commensurability.[9] Looking at publication only within the global top decile of citations received increases international commensurability.

I.E Topical Alignment with Government Priorities

Metric I.E.1 is a measure of topical alignment to DoD priorities. Based on the public sources discussed in Chapter Three, the quantum communications subfields of low DoD priority are QKD and quantum cryptography. The quantum sensing subfield of low DoD priority is quantum imaging (which includes quantum illumination and quantum radar but not passive electromagnetic quantum sensors). This metric is calculated as the proportion of a country's total publications (Metric A.1) that fall into subfields designed to be of low priority to DoD.

I.F Degree of Domestic and International Collaboration

Metric I.F.1 is a measure of the connectedness of the domestic collaboration network for the focal research field. It is calculated as two times the number of collaborations as a proportion of the number of research units in the networks. In the language of network theory, this metric is *average weighted network degree* and is calculated as $2m/n$, where m is the number of edges and n is the number of nodes. This metric can be interpreted as the average number of domestic collaborators a domestic research unit has within the focal scientific field.

Metric I.F.2 is a measure of international research collaboration. It is measured as the proportion of a country's total publications that feature at least one coauthor with an international affiliation.

I.G Risk of Technology Leakage

Metric I.G.1 is measured as the proportion of a country's total publications (Metric I.A.1) that feature a coauthor with an affiliation based in a strategic competitor nation.

Metric I.G.2 is a measure of technology leakage risk to a strategic competitor nation's military. For our study, it is measured as the number of unique U.S.-based authors to have coauthored at least one publication within a given application domain with an author affiliated with a Chinese military-affiliated university. The Chinese military-affiliated universities used to calculate this measure were Naval Engineering University, National University of Defense Technology, Air Force Engineering University, PLA Information Engineering University, PLA University of Science & Technology, Academy of Military Medical Sciences, Academy of Armored Forces Engineering, PLA Air Force Aviation University, and Naval Aviation University.

---

[8] Jon Schmid and Fei-Ling Wang, "Beyond National Innovation Systems: Incentives and China's Innovation Performance," *Journal of Contemporary China*, Vol. 26, No. 104, 2017.

[9] Jon Schmid, *An Open-Source Method for Assessing National Scientific and Technological Standing: With Applications to Artificial Intelligence and Machine Learning*, Santa Monica, Calif.: RAND Corporation, RR-A1482-3, 2021.

Government Metrics Methodology

Compared with other data fields within our publications data set, the funding field—a free-text field—is relatively messy. For example, in our data, an NSF-funded publication may appear as, among others, "NSF—Directorate for Mathematical & Physical Sciences (MPS)," "U.S. National Science Foundation," "National Science Foundation (NSF)," or "NSF—Directorate for Computer & Information Science & Engineering (CISE)." In order to ensure that each of the publications are included under the total for the NSF, we apply a two-step process to clean the funding agency text field.

First, we apply a list-cleanup algorithm. To this end, we use Vantage Point, a text-mining software. Vantage Point's list-cleanup function searches text fields to identify like text entries that appear in different ways. In order for a field to be combined, the function requires at least a 68-percent two-way and a 51-percent one-way match between names based on the number of parts of the name. The algorithm ignores stemming text and common terms.

Second, following the application of the cleanup function, the research team conducted a hand search for the top-20 funding organizations for the United States and China. The hand search consisted of conducting a variety of searches for known synonyms for the funding agencies. For example, to find alternative variants of projects funded by the U.S. Air Force, we searched terms such as "USAF," "United States Air Force," as well as the Air Force research laboratories. This text-cleaning process was applied for all three application domains. When grouping funding agencies, we seek to select the parent organization. For example, research funded through the Physics Frontier Center at the Joint Quantum Institute, Center for Ultra-cold Atoms (CUA), and Institute for Quantum Information and Matter (Caltech) are funded through an NSF grant, so these publications are assigned to the NSF bin.

The HHI for funding sources is calculated as the sum of the squared funded publication shares for funding sources within a country's research funding system. More formally,

$$HHI\ Funding\ agencies = \sum_{1}^{N} s_i^2 \text{ where } s_i \text{ is the funded publication share of the ith funding agency.}$$

Funded publication share refers to the ratio of the number of publications on which a given funding agency is listed as a funder and the number of total domestic funding slots for a given application domain. We limit the calculation to include only each country's top-20 funding sources. For example, for the quantum sensing application domain, the NSF is listed as funder on 381 publications, and there are 1,610 total quantum sensing funding slots. Therefore, the NSF has a funding share of 24 percent (381/1,610 = 0.24) for quantum sensing. This proportion is calculated for the top-20 funding sources. The sum of the squared funded publication shares is the HHI for the application domain in question. The higher a country's HHI for funding sources, the more concentrated is its research funding.

Industry Metrics Methodology

Distribution of QIB Firms by VC Funding

In this subsection, we describe our methodology for creating Figures 3.10 and 3.11 in Chapter Three.

For Figure 3.10, we started with the 50 companies (all start-ups) in the QED-C that we judged to be primarily or exclusively focused on quantum technology. We searched the financial reporting website Crunchbase.com for each of these 50 companies and were able to find announced VC funding levels for 20 of them. We verified these VC funding numbers with a second source (Golden.com). The VC funding numbers are current as of June 21, 2021.

In Figure 3.11, each bin roughly corresponds to a different VC funding round. Technology start-ups, both conventional and quantum, begin with a "seed round" of VC funding and then progress to subsequent rounds of funding if investors are satisfied with the progress that has been made since the previous round. At each round, some start-ups are not able to convince investors to give them more money and fold, while others are bought by larger existing companies for their intellectual property or talent pool.

We found a data source that gave information (current as of September 2018) about VC funding by a large sample of over 1,100 technology start-ups that raised a seed round in 2008–2010.[10] Unfortunately, this data source did not directly give the full distribution of the start-ups' VC funding levels but only summary statistics aggregated by seed round (e.g., statistics for the companies in their second VC round, for those in their third VC round, etc.). So we were forced to use the seed-round number as a proxy for total VC in order to roughly estimate the distribution of VC levels across companies.

We made the simplifying assumption that every company that raised a third round received more money than every company that raised a second round, and so forth. Under this assumption, the VC round numbers correspond to nonoverlapping bins in the funding distribution: the first bin corresponds to the companies that raised one round, the second bin corresponds to companies that raised two, and so on.

The median levels of cumulative funding raised by the firms in each round are displayed as red dots in Figure A.3. We set the bin thresholds (black bars) at the midpoints between funding round medians (rounded to the nearest million), which gave us the bins for our final distribution.

By making further assumptions, we were able to use the summary data from our data source to estimate the proportion of start-ups in each round at a given point in time[11] and therefore generate the approximate distribution of firms displayed by the blue bars in

---

[10] CB Insights, 2018. One caveat is that the time periods covered for the quantum companies and the all-technology sector are overlapping but not identical: the all-technology sector data cover 2008–2018, while the quantum sector data are through 2021 and include companies with a range of founding dates.

[11] For each round, we had the percentage of companies that would successfully raise another round. We also had the average of how long it would take them to do so. We assumed that each start-up took the aver-

FIGURE A.3
Illustration of the Methodology for Generating the Bins

Figure A.3. We independently collected the total amount of money that each quantum start-up raised and calculated the distribution (the orange bars in Figure 3.11) using the same bins as for all the technology companies.

Foreign Supply-Chain Dependencies

We analyzed the qualitative data obtained from our conversations with industry using the Dedoose analytic software, which allowed us to "code" sections of our notes pertaining to various subjects of interest—for example, foreign supply-chain dependencies, single-source components, or critical materials.[12] We then used the software to dynamically group excerpts as needed to answer the following questions:

- Which components come from foreign suppliers, and why?
- Which countries do these components come from, and what is their relationship with the United States?
- Which foreign companies are particularly integral to the quantum supply chain?
- For which components are there a very limited number of suppliers, and to what degree do these overlap with those coming from foreign sources?
- Which materials and components are critical to various quantum technologies, and do they face supply-chain risk due to foreign or single-source supply-chain dependencies?
- What are potential implications of supply-chain dependencies?

One potential limitation of our approach is that we agreed not to discuss any business proprietary information so that this analysis could be shared publicly.

---

age time and either stopped operating independently or raised another round at the end of the period. This let us calculate how many start-ups to expect in each round of funding at a specific point in time.

[12] Dedoose Version 9.0.15, web application for managing, analyzing, and presenting qualitative and mixed method research data, 2021. Los Angeles, Calif.: SocioCultural Research Consultants, LLC.

Technical Metrics Methodology

In this section, we describe the patent analysis methodology.

Inventors submit patent applications to national and international patent-granting organizations with the expectation of legal protection for the invention described in the application, should the patent be granted, for 20 years in the country or countries in which the application is filed. The resources invested in developing the (invention and) patent application are akin to a bet that the application will result in a patent grant that protects a market innovation providing a greater return than that investment. Moreover, patent applications are customarily filed first in the country in which the invention is made.[13] Accordingly, the volume of patent applications and/or issued patents in a technical area provides a measure of the potential innovations in that technical area in the country in which they are initially filed.[14] Moreover, the number of unique entities to which those applications or patents are assigned indicates the breadth of organizations driving those potential innovations.

Our patent data is derived from the IFI CLAIMS Direct platform which includes full-text patent data from 38 countries, together with metadata such as filing date, patent classes, assignees, and drawings. Patent text is machine translated to English and format is standardized to facilitate analysis.[15] For each technical sector, we count all patent applications that have any of the keywords associated with that sector anywhere in its text. For quantum computing, which has a specific technology subclassification under the Cooperative Patent Classification (CPC) Scheme used by many national and international patent-granting organizations (G06N 10/00), we include patent applications assigned to this subclassification that were not captured by the keywords. We record the year in which each patent application was filed as well as the priority year of a family of patent applications when multiple applications are filed on a single invention. For issued patents, we record the year of the patent grant.

The time dependence of the cumulative number of patent applications in technical areas defined by patent-granting organizations (e.g., according to the CPC Scheme) generally follows a logistic or S-curve when strong interest in these areas arises.[16] This same S-curve behavior also occurs for patent applications containing certain words or groups of words.[17]

---

[13] In the United States, a Foreign Filing License is required from the Department of Commerce to file a patent application overseas. This practice is common among national patent-granting organizations.

[14] Excluding those innovations stemming from inventions for which a patent application is not filed—for example, those kept as trade secrets.

[15] This data set includes more than 100 sources and 125 million records. For a detailed description, see IFI CLAIMS Patent Services, "CLAIMS Direct Data Collection," *Ificlaims.com*, webpage, 2021.

[16] As evidenced by a rapid increase in the number of patent applications assigned to a specific technology subclassification. This S-curve also represents diffusion of the technology into new application areas. See Eusebi and Silberglitt, 2014.

[17] For the example of meta-materials, see Richard Silberglitt, *New and Critical Materials: Identifying Potential Dual-Use Areas*, Santa Monica, Calif.: RAND Corporation, CT-513, 2019.

We observe such S-curve dependence for the cumulative number of patent applications that contain any of the keywords that define quantum computing, quantum communications, and quantum sensing.

When we observe S-curves, either for technology subclassifications or for keywords that define a technology sector, we record the position of each assignee's patent applications along these curves. Early positions indicate patent applications filed early in the development of the technology, when there were few patent filings, and thus greater innovation potential.[18] When we observe S-curves for individual keywords or keyword phrases, we note early patent applications and assignees in these cases as well.

---

[18] Independent analysis of litigation awards suggests that early patent applications also represent the potentially most valuable intellectual property in their technical area.

Abbreviations

AI	artificial intelligence
CAGR	compound annual growth rate
COTS	commercial off-the-shelf
DARPA	Defense Advanced Research Projects Agency
DoD	(United States) Department of Defense
DOE	(United States) Department of Energy
DSB	(U.S. Department of Defense) Defense Science Board
FY	fiscal year
FYP	(People's Republic of China) Five-Year Plan
GDP	gross domestic product
HFNL	Hefei National Laboratory for Physical Sciences at the Microscale
HHI	Herfindahl-Hirschman Index
IARPA	Intelligence Advanced Research Projects Activity
ISR	intelligence, surveillance, and reconnaissance
NASA	National Aeronautics and Space Administration
NISQ	noisy intermediate-scale quantum
NIST	National Institutes of Standards and Technology
NSA	National Security Agency
NSF	National Science Foundation
NSFC	National Natural Science Foundation of China
OSD	Office of the Secretary of Defense
PNT	positioning, navigation, and timing
PRC	People's Republic of China
QEC	quantum error correction
QED-C	Quantum Economic Development Consortium
QIB	quantum industrial base
QIS	quantum information science
QKD	quantum key distribution
R&D	research and development
SBIR	Small Business Innovation Research
SCT	superconducting transmon
SME	subject-matter expert
STTR	Small Business Technology Transfer
VC	venture capital

References

Aaronson, Scott, *Quantum Computing Since Democritus*, Cambridge: Cambridge University Press, 2013.

———, "Read the Fine Print," *Nature Physics*, Vol. 11, 2015, pp. 291–293.

Aaronson, Scott, and Alex Arkhipov, "The Computational Complexity of Linear Optics," *Theory of Computing*, Vol. 9, No. 4, February 2013, pp. 143–252.

Aksnes, Dag W., Liv Langfeldt, and Paul Wouters, "Citations, Citation Indicators, and Research Quality: An Overview of Basic Concepts and Theories," *SAGE Open*, January 2019.

Arute, F., K. Arya, R. Babbush, D. Bacon, J. C. Bardin, R. Barends, R. Biswas, S. Boixo, F. G. S. L. Brandao, D. A. Buell, B. Burkett, Y. Chen, Z. Chen, B. Chiaro, R. Collins, W. Courtney, A. Dunsworth, E. Farhi, B. Foxen, A. Fowler, C. Gidney, M. Giustina, R. Graff, K. Guerin, S. Habegger, M. P. Harrigan, M. J. Hartmann, A. Ho, M. Hoffmann, T. Huang, T. S. Humble, S. V. Isakov, E. Jeffrey, Z. Jiang, D. Kafri, K. Kechedzhi, J. Kelly, P. V. Klimov, S. Knysh, A. Korotkov, F. Kostritsa, D. Landhuis, M. Lindmark, E. Lucero, D. Lyakh, S. Mandrà, J. R. McClean, M. McEwen, A. Megrant, X. Mi, K. Michielsen, M. Mohseni, J. Mutus, O. Naaman, M. Neeley, C. Neill, M. Y. Niu, E. Ostby, A. Petukhov, J. C. Platt, C. Quintana, E. G. Rieffel, P. Roushan, N. C. Rubin, D. Sank, K. J. Satzinger, V. Smelyanskiy, K. J. Sung, M. D. Trevithick, A. Vainsencher, B. Villalonga, T. White, Z. J. Yao, P. Yeh, A. Zalcman, H. Neven, and J. M. Martinis, "Quantum Supremacy Using a Programmable Superconducting Processor," *Nature*, Vol. 574, 2019, pp. 505–510.

Barzanjeh, Shabir, Saikat Guha, Christian Weedbrook, David Vitali, Jeffrey H. Shapiro, and Stefano Pirandola, "Microwave Quantum Illumination," *Physical Review Letters*, Vol. 114, 2015, p. 080503.

Beall, Jeffrey, "Essential Information About Predatory Publishers and Journals," *International Higher Education*, Vol. 86, 2016, pp. 2–3.

Biedermann, G. W., H. J. McGuinness, A. V. Rakholia, Y.-Y. Jau, D. R. Wheeler, J. D. Sterk, and G. R. Burns, "Atom Interferometry in a Warm Vapor," *Physical Review Letters*, Vol. 118, 2017.

Bourzac, Katerine, "Chemistry Is Quantum Computing's Killer App," *Chemical & Engineering News*, Vol. 95, No. 43, October 30, 2017.

CB Insights, "Venture Capital Funnel Shows Odds of Becoming a Unicorn Are About 1%," *CB Insights.com*, September 6, 2018. As of August 5, 2021: https://www.cbinsights.com/research/venture-capital-funnel-2/

Chang, W., C. Li, Y.-K. Wu, N. Jiang, S. Zhang, Y.-F. Pu, X.-Y. Chang, and L. M. Duan, "Long-Distance Entanglement Between a Multiplexed Quantum Memory and a Telecom Photon," *Physics Review*, Vol. 9, November 14, 2019, p. 041033. As of August 5, 2021: https://journals.aps.org/prx/pdf/10.1103/PhysRevX.9.041033

Chen, Y., M. Zopf, R. Keil, F. Ding, and O. G. Schmidt, "Highly-Efficient Extraction of Entangled Photons from Quantum Dots Using a Broadband Optical Antenna," *Nature Communications*, Vol. 9, No. 1, 2018, p. 2994.

Chen, Yu, "Developing Technologies Towards an Error-Corrected Quantum Computer," speech delivered at IEEE Quantum Week 2020, October 13, 2020.

Chen, Yu-Ao, Qiang Zhang, and Jian-Wei Pan, "An Integrated Space-to-Ground Quantum Communication Network over 4,600 Kilometres," *Nature*, Vol. 589, 2021, pp. 214–219.

Chen, Yulei, Zhonghao Li, Hao Guo, Dajin Wu, and Jun Tang, "Simultaneous Imaging of Magnetic Field and Temperature Using a Wide-Field Quantum Diamond Microscope," *European Journal of Physics Quantum Technology*, Vol. 8, No. 8, 2021.

ColdQuanta, Inc., "Cold Atom Quantum Computing," video, YouTube, November 23, 2020. As of August 5, 2021:
https://www.youtube.com/watch?v=M0LO19PpDTw

Cornillie, Chris, "Finding Artificial Intelligence Money in the Fiscal 2020 Budget," *Bloomberg Government*, March 28, 2019. As of August 5, 2021:
https://about.bgov.com/news/finding-artificial-intelligence-money-fiscal-2020-budget/

Defense Science Board, "Applications of Quantum Technologies—Executive Summary," October 2019. As of August 5, 2021:
https://dsb.cto.mil/reports/2010s/DSB_QuantumTechnologies_Executive%20Summary_10.23.2019_SR.pdf

Denchev, Vasil S., Sergio Boixo, Sergei V. Isakov, Nan Ding, Ryan Babbush, Vadim Smelyanskiy, John Martinis, and Hartmut Neven, "What Is the Computational Value of Finite-Range Tunneling?" *Physical Review X*, Vol. 6, No. 3, 2016, p. 031015.

Dowling, Jonathan P., and J. Milburn Gerard, "Quantum Technology: The Second Quantum Revolution," *Philosophical Transactions of the Royal Society A*, Vol. 361, 2003, pp. 1655–1674.

Du, Dounan, Paul Stankus, Olli-Pentti Saira, Mael Flament, Steven Sagona-Stophel, Mehdi Namazi, Dimitrios Katramatos, and Eden Figueroa, "An Elementary 158 Km Long Quantum Network Connecting Room Temperature Quantum Memories," *ArXiv.org*, *Quantum Physics*, January 2021. As of August 5, 2021:
https://arxiv.org/abs/2101.12742

Ebadi, Sepehr, Tout T. Tang, Harry Levine, Alexander Keesling, Giulia Semeghini, Ahmed Omran, Doley Bluvstein, Rhine Samajdar, Hannes Pichler, Wen Wei Ho, Soonwon Choi, Subir Sachdev, Markus Greiner, Vladan Vuletić, and Mikhail D. Lukin, "Quantum Phases of Matter on a 256-Atom Programmable Quantum Simulator," *Nature*, Vol. 595, No. 7866, July 2021, pp. 227–232.

Egan, Laird, Dripto M. Debroy, Crystal Noel, Andrew Risinger, Daiwei Zhu, Debopriyo Biswas, Michael Newman, Muyuan Li, Kenneth R. Brown, Marko Cetina, and Christopher Monroe, "Fault-Tolerant Control of an Error-Corrected Qubit," *Nature*, Vol. 598, October 2021, pp. 281–286.

Etzkowitz, Henry, and Chunyan Zhou, *The Triple Helix: University-Industry-Government Innovation and Entrepreneurship*, London: Routledge, 2017.

Eusebi, Christopher A., and Richard Silberglitt, *Identification and Analysis of Technological Emergence Using Patent Classification*, Santa Monica, Calif.: RAND Corporation, RR-629-OSD, 2014. As of August 5, 2021:
https://www.rand.org/pubs/research_reports/RR629.html

Fei, Y. Y., X. D. Meng, M. Gao, H. Wang, and Z. Ma, "Quantum Man-in-the-Middle Attack on the Calibration Process of Quantum Key Distribution," *International Journal of Scientific Reports*, Vol. 8, 2018, p. 4283.

Feynman, Richard P., "Simulating Physics with Computers," *International Journal of Theoretical Physics*, Vol. 21, 1982, pp. 467–488.

Figliola, Patricia Moloney, *Federal Quantum Information Science: An Overview*, Washington, D.C.: Congressional Research Service, IF10872, July 2, 2018.

Foxen, B., C. Neill, A. Dunsworth, P. Roushan, B. Chiaro, A. Megrant, J. Kelly, Z. Chen, K. Satzinger, R. Barends, F. Arute, K. Arya, R. Babbush, D. Bacon, J. C. Bardin, S. Boixo, D. Buell, B. Burkett, Y. Chen, R. Collins, E. Farhi, A. Fowler, C. Gidney, M. Giustina, R. Graff, M. Harrigan, T. Huang, S. V. Isakov, E. Jeffrey, Z. Jiang, D. Kafri, K. Kechedzhi, P. Klimov, A. Korotkov, F. Kostritsa, D. Landhuis, E. Lucero, J. McClean, M. McEwen, X. Mi, M. Mohseni, J. Y. Mutus, O. Naaman, M. Neeley, M. Niu, A. Petukhov, C. Quintana, N. Rubin, D. Sank, V. Smelyanskiy, A. Vainsencher, T. C. White, Z. Yao, P. Yeh, A. Zalcman, H. Neven, and J. M. Martinis, "Demonstrating a Continuous Set of Two-Qubit Gates for Near-Term Quantum Algorithms," *Physical Review Letters*, Vol. 125, No. 12, September 15, 2020, p. 120504.

Gambetta, Jay, "IBM's Roadmap for Scaling Quantum Technology," *IBM Research Blog*, September 15, 2020. As of August 5, 2021:
https://www.ibm.com/blogs/research/2020/09/ibm-quantum-roadmap/

Gerbert, Philipp, and Frank Ruess, *The Next Decade in Quantum Computing—and How to Play*, Boston, Boston Consulting Group, November 2018.

Gibney, Elizabeth, "Quantum Gold Rush: The Private Funding Pouring into Quantum Start-Ups," *Nature News Feature*, October 2, 2019a. As of August 5, 2021:
https://www.nature.com/articles/d41586-019-02935-4

―――, "Hello Quantum World! Google Published Landmark Quantum Supremacy Claim," *Nature*, October 23, 2019b.

Giles, Martin, "The Man Turning China into a Quantum Superpower," *MIT Technology Review*, December 19, 2018. As of August 5, 2021:
https://www.technologyreview.com/s/612596/
the-man-turning-china-into-a-quantum-superpower/

Gong, Ming, Shiyu Wang, Chen Zha, Ming-Cheng Chen, He-Liang Huang, Yulin Wu, Qingling Zhu, Youwei Zhao, Shaowei Li, Shaojun Guo, Haoran Qian, Yangsen Ye, Fusheng Chen, Chong Ying, Jiale Yu, Daojin Fan, Dachao Wu, Hong Su, Hui Deng, Hao Rong, Kaili Zhang, Sirui Cao, Jin Lin, Yu Xu, Lihua Sun, Cheng Guo, Na Li, Futian Liang, V. M. Bastidas, Kae Nemoto, W. J. Munro, Yong-Heng Huo, Chao-Yang Lu, Cheng-Zhi Peng, Xiaobo Zhu, and Jian-Wei Pan, "Quantum Walks on a Programmable Two-Dimensional 62-Qubit Superconducting Processor," *Science*, Vol. 372, No. 6545, May 28, 2021, pp. 948–952.

Gonzales, Daniel, Sarah Harting, Mary Kate Adgie, Julia Brackup, Lindsey Polley, and Karlyn D. Stanley, *Unclassified and Secure: A Defense Industrial Base Cyber Protection Program for Unclassified Defense Networks*, Santa Monica, Calif.: RAND Corporation, RR-4227-RC, 2020. As of August 5, 2021:
https://www.rand.org/pubs/research_reports/RR4227.html

Google Quantum AI, "Exponential Suppression of Bit or Phase Errors with Cyclic Error Correction," *Nature*, Vol. 595, 2021, pp. 383–387.

Grover, Lov K., "A Fast Quantum Mechanical Algorithm for Database Search," *28th Annual ACM Symposium on the Theory of Computing*, 1996.

GWR Instruments, Inc., "iGRAV® Gravity Sensors," webpage, 2019. As of August 5, 2021:
https://www.gwrinstruments.com/igrav-gravity-sensors.html

Hackett, Robert, "IBM Plans a Huge Leap in Superfast Quantum Computing by 2023," *Fortune*, September 15, 2020. As of August 5, 2021:
https://fortune.com/2020/09/15/ibm-quantum-computer-1-million-qubits-by-2030/

Harrow, Aram W., Avinatan Hassidim, and Seth Lloyd, "Quantum Algorithm for Linear Systems of Equations," *Physical Review Letters*, Vol. 103, No. 15, 2009.

Herman, Arthur, "At Last America Is Moving on Quantum," *Forbes*, August 20, 2018. As of August 5, 2021:
https://www.forbes.com/sites/arthurherman/2018/08/20/at-last-america-is-moving-on -quantum/

Honeywell, "Get to Know Honeywell's Latest Quantum Computer System Model H1: Technical Details of Our Highest Performing System," webpage, undated. As of August 5, 2021:
https://www.honeywell.com/us/en/news/2020/10/get-to-know-honeywell-s-latest-quantum -computer-system-model-h1

Honeywell, "Honeywell Sets New Record for Quantum Computing Performance," March 2021. As of August 5, 2021:
https://www.honeywell.com/us/en/news/2021/03/honeywell-sets-new-record-for-quantum -computing-performance

Huanqiu Shibao (Global Times, Chinese edition), "Quantum Technology China Presses the 'Fast Forward Button' 4,200 Related Enterprises in the Quantum Field in My Country," October 22, 2020. As of August 5, 2021:
https://baijiahao.baidu.com/s?id=1681243805758684852&wfr=spider&for=pc

Huaxia, ed., "Xi Stresses Advancing Development of Quantum Science and Technology," *Xinhuanet.com*, October 17, 2020. As of August 5, 2021:
http://www.xinhuanet.com/english/2020-10/17/c_139447976.htm

IFI CLAIMS Patent Services, "CLAIMS Direct Data Collection," *Ificlaims.com*, webpage, 2021. As of August 5, 2021:
https://www.ificlaims.com/product/product-data-collection.htm#:~:text=CLAIMS%20 Direct%20from%20IFI%20CLAIMS%20integrates%20patent%20data,text%20translations%20 and%20other%20data%20enhancements.%20Skip%20navigation

Jurcevic, Petar, Ali Javadi-Abhari, Lev S. Bishop, Isaac Lauer, Daniela F. Bogorin, Markus Brink, Lauren Capelluto, Oktay Günlük, Toshinari Itoko, Naoki Kanazawa, Abhinav Kandala, George A. Keefe, Kevin Krsulich, William Landers, Eric P. Lewandowski, Douglas T. McClure, Giacomo Nannicini, Adinath Narasgond, Hasan M. Nayfeh, Emily Pritchett, Mary Beth Rothwell, Srikanth Srinivasan, Neereja Sundaresan, Cindy Wang, Ken X. Wei, Christopher J. Wood, Jeng-Bang Yau, Eric J. Zhang, Oliver E. Dial, Jerry M. Chow, and Jay M. Gambetta, "Demonstration of Quantum Volume 64 on a Superconducting Quantum Computing System," *ArXiv.org*, *Quantum Physics*, 2008. As of August 5, 2021:
http://arxiv.org/abs/2008.08571

Kandala, Abhinav, Antonio Mezzacapo, Kristan Temme, Maika Takita, Markus Brink, Jerry M. Chow, and Jay M. Gambetta, "Hardware-Efficient Variational Quantum Eigensolver for Small Molecules and Quantum Magnets," *Nature*, Vol. 549, 2017, pp. 242–246.

Kania, Elsa B., and John K. Costello, *Quantum Hegemony? China's Ambitions and the Challenge to U.S. Innovation Leadership*, Washington, D.C., Center for a New American Security, September 12, 2018. As of August 5, 2021:
https://www.cnas.org/publications/reports/quantum-hegemony

Kim, Donggyu, Mohamed I. Ibrahim, Christopher Foy, Matthew E. Trusheim, Ruonan Han, and Dirk R. Englund, "A CMOS-Integrated Quantum Sensor Based on Nitrogen-Vacancy Centres," *Nature Electronics*, Vol. 2, 2019, pp. 284–289.

Krasnyak, Olga, "Science Diplomacy and Soviet-American Academic and Technical Exchanges," *The Hague Journal of Diplomacy*, Vol. 15, No. 3, 2020, pp. 398–408.

Krutyanskiy, V., M. Meraner, J. Schupp, V. Krcmarsky, H. Hainzer, and B. P. Lanyon, "Light-Matter Entanglement over 50 Km of Optical Fibre," *Nature Quantum Information*, Vol. 5, 2019, p. 72.

Ladd, T. D., F. Jelezko, R. Laflamme, Y. Nakamura, C. Monroe, and J. L. O'Brien, "Quantum Computers," *Nature*, Vol. 464, 2010, pp. 45–53.

Lardinois, Frederic, "Rigetti Computing Goes Public via SPAC Merger," *TechCrunch*, October 6, 2021. As of August 5, 2021:
https://techcrunch.com/2021/10/06/rigetti-computing-goes-public-via-spac-merger/

Lee, Patty, "Unleashing the Power of Quantum for Everyone," speech delivered at the Future Compute Virtual Conference, held virtually, February 10, 2021.

Li, Ruoyu, Luca Petit, David P. Franke, Juan Pablo Dehollain, Jonas Helsen, Mark Steudtner, Nicole K. Thomas, Zachary R. Yoscovits, Kanwal J. Singh, Stephanie Wehner, Lieven M. K. Vandersypen, James S. Clarke, and Menno Veldhorst, "A Crossbar Network for Silicon Quantum Dot Qubits | Science Advances," *Science Advances Magazine*, Vol. 4, No. 7, July 6, 2018, p. eaar3960. As of August 5, 2021:
https://advances.sciencemag.org/content/4/7/eaar3960

Limes, M. E., E. L. Foley, T. W. Kornack, S. Caliga, S. McBride, A. Braun, W. Lee, V. G. Lucivero, and M. V. Romalis, "Portable Magnetometry for Detection of Biomagnetism in Ambient Environments," *Physical Review Applied*, Vol. 14, 2020, p. 011002.

Liu, J., R. Su, Y. Wei, B. Yao, S. F. Covre da Silva, Y. Yu, J. Iles-Smith, K. Srinivasan, A. Rastelli, J. Li, and X. Wang, "A Solid-State Source of Strongly Entangled Photon Pairs with High Brightness and Indistinguishability," *Nature Nanotechnology*, Vol. 14, 2019, pp. 586–593.

Lohrmann, Alexander, Chithrabhanu Perumangatt, Aitor Villar, and Alexander Ling, "Broadband Pumped Polarization Entangled Photon-Pair Source in a Linear Beam Displacement Interferometer," *Applied Physics Letters*, Vol. 116, 2020, p. 021101.

Ma, Zhaohui, Jia-Yang Chen, Zhan Li, Chao Tang, Yong Meng Sua, Heng Fan, and Yu-Ping Huang, "Ultrabright Quantum Photon Sources on Chip," *Physical Review Letters*, Vol. 125, December 2020, p. 263602.

Mattis, Jim. *Summary of the National Defense Strategy of the United States of America: Sharpening the American Military's Competitive Edge*, Washington, D.C., United States Department of Defense, 2018. As of August 5, 2021:
https://dod.defense.gov/Portals/1/Documents/pubs/2018-National-Defense-Strategy-Summary.pdf

Ménoret, Vincent, Pierre Vermeulen, Nicolas Le Moigne, Sylvain Bonvalot, Philippe Bouyer, Arnaud Landragin, and Bruno Desruelle, "Gravity Measurements Below 10^{-9} g with a Transportable Absolute Quantum Gravimeter," *Nature Scientific Reports*, Vol. 8, 2018, p. 12300.

Merton, R. K., "Foreword," in E. Garfield, ed., *Citation Indexing—Its Theory and Application in Science, Technology, and Humanities*, New York: Wiley, 1979, pp. v–ix.

Meyer-Scott, Evan, Nidhin Prasannan, Christof Eigner, Viktor Quiring, John M. Donohue, Sonja Barkhofen, and Christine Silberhorn, "High-Performance Source of Spectrally Pure, Polarization Entangled Photon Pairs Based on Hybrid Integrated-Bulk Optics," *Optics Express*, Vol. 26, 2018, pp. 32475–32490.

Miller, Ron, "IBM Makes 20 Qubit Quantum Computing Machine Available as a Cloud Service," *Tech Crunch*, November 10, 2017. As of August 5, 2021:
https://techcrunch.com/2017/11/10/ibm-passes-major-milestone-with-20-and-50-qubit-quantum-computers-as-a-service/

Moor Insights and Strategy and Paul Smith-Goodson, "IonQ Releases a New 32-Qubit Trapped-Ion Quantum Computer with Massive Quantum Volume Claims," *Forbes*, October 7, 2020. As of August 5, 2021:
https://www.forbes.com/sites/moorinsights/2020/10/07/ionq-releases-a-new-32-qubit-trapped-ion-quantum-computer-with-massive-quantum-volume-claims/?sh=444332823b39

National Academies of Sciences, Engineering, and Medicine, *Quantum Computing: Progress and Prospects*, Washington, D.C.: National Academies Press, 2019.

"National Quantum Initiative Supplement to the President's FY 2021 Budget," Subcommittee on Quantum Information Science, Committee on Science of the National Science and Technology Council, January 2021.

National Science and Technology Council, Committee on Homeland and National Security, Subcommittee on Economic and Security Implications of Quantum Science, *The Role of International Talent in Quantum Information Science*, Washington, D.C.: National Science and Technology Council, October 2021.

National Security Agency, "Quantum Key Distribution (QKD) and Quantum Cryptography (QC)," undated. As of January 14, 2022:
https://www.nsa.gov/Cybersecurity/Quantum-Key-Distribution-QKD-and-Quantum-Cryptography-QC/

Neill, C., T. McCourt, X. Mi, Z. Jiang, M. Y. Niu, W. Mruczkiewicz, I. Aleiner, F. Arute, K. Arya, J. Atalaya, R. Babbush, J. C. Bardin, R. Barends, A. Bengtsson, A. Bourassa, M. Broughton, B. B. Buckley, D. A. Buell, B. Burkett, N. Bushnell, J. Campero, Z. Chen, B. Chiaro, R. Collins, W. Courtney, S. Demura, A. R. Derk, A. Dunsworth, D. Eppens, C. Erickson, E. Farhi, A. G. Fowler, B. Foxen, C. Gidney, M. Giustina, J. A. Gross, M. P. Harrigan, S. D. Harrington, J. Hilton, A. Ho, S. Hong, T. Huang, W. J. Huggins, S. V. Isakov, M. Jacob-Mitos, E. Jeffrey, C. Jones, D. Kafri, K. Kechedzhi, J. Kelly, S. Kim, P. V. Klimov, A. N. Korotkov, F. Kostritsa, D. Landhuis, P. Laptev, E. Lucero, O. Martin, J. R. McClean, M. McEwen, A. Megrant, K. C. Miao, M. Mohseni, J. Mutus, O. Naaman, M. Neeley, M. Newman, T. E. O'Brien, A. Opremcak, E. Ostby, B. Pató, A. Petukhov, C. Quintana, N. Redd, N. C. Rubin, D. Sank, K. J. Satzinger, V. Shvarts, D. Strain, M. Szalay, M. D. Trevithick, B. Villalonga, T. C. White, Z. Yao, P. Yeh, A. Zalcman, H. Neven, S. Boixo, L. B. Ioffe, P. Roushan, Y. Chen, and V. Smelyanskiy, "Accurately Computing the Electronic Properties of a Quantum Ring," *Nature*, Vol. 594, 2021, pp. 508–512.

Office of the Secretary of Defense for Acquisition and Sustainment (OSD A&S) Industrial Policy, *Fiscal Year 2019 Industrial Capabilities Report to Congress*, Washington, D.C.: OSD A&S, June 23, 2020.

Office of the Secretary of Defense for Acquisition and Sustainment (OSD A&S) Industrial Policy, *Fiscal Year 2020 Industrial Capabilities Report to Congress*, January 2021. As of August 5, 2021:
https://media.defense.gov/2021/Jan/14/2002565311/-1/-1/0/FY20-INDUSTRIAL-CAPABILITIES-REPORT.PDF

Pan, Jian-Wei, "Improve the Development of Our Nation's Quantum Technology," *Red Flag Manuscripts*, in Chinese, December 7, 2020. As of August 5, 2021:
http://news.ustc.edu.cn/info/1056/73476.htm

Pang, Xiao-Ling, Ai-Lin Yang, Chao-Ni Zhang, Jian-Peng Dou, Hang Li, Jun Gao, and Xian-Min Jin, "Hacking Quantum Key Distribution via Injection Locking," *Physics Review Applied*, Vol. 13, 2020, p. 034008.

Pino, J. M., J. M. Dreiling, C. Figgatt, J. P. Gaebler, S. A. Moses, M. S. Allman, C. H. Baldwin, M. Foss-Feig, D. Hayes, K. Mayer, C. Ryan-Anderson, and B. Neyenhuis, "Demonstration of the Trapped-Ion Quantum CCD Computer Architecture," *Nature*, Vol. 592, No. 7853, April 2021, pp. 209–213.

Reiher, Markus, Nathan Wiebe, Krysta Marie Svore, Dave Wrecker, and Matthias Troyer, "Elucidating Reaction Mechanisms on Quantum Computers," *Proceedings of the National Academy of Sciences*, Vol. 114, No. 29, 2017, pp. 7555–7560.

Rigetti, "What," webpage, undated. As of August 5, 2021:
https://www.rigetti.com/what

Rinia, E. J., T. N. van Leeuwen, H. G. van Vuren, and A. F. J. van Raan, "Comparative Analysis of a Set of Bibliometric Indicators and Central Peer Review Criteria: Evaluation of Condensed Matter Physics in the Netherlands," *Research Policy*, Vol. 27, 1998, pp. 95–107.

Salge, Torsten Oliver, Erk Peter Piening, and Nils Foege, "Exploring the Dark Side of Innovation Collaboration: A Resource-Based Perspective," *Academy of Management Proceedings*, Vol. 2013, No. 1, p. 12061.

Savitz, Scott, Miriam Matthews, and Sarah Weilant, *Assessing Impact to Inform Decisions: A Toolkit on Measures for Policymakers*, Santa Monica, Calif.: RAND Corporation, TL-263-OSD, 2017. As of August 5, 2021:
https://www.rand.org/pubs/tools/TL263.html

Schmid, Jon, *An Open-Source Method for Assessing National Scientific and Technological Standing: With Applications to Artificial Intelligence and Machine Learning*, Santa Monica, Calif.: RAND Corporation, RR-A1482-3, 2021. As of August 5, 2021:
https://www.rand.org/pubs/research_reports/RRA1482-3.html

Schmid, Jon, Sergey A. Kolesnikov, and Jan Youtie, "Plans Versus Experiences in Transitioning Transnational Education into Research and Economic Development: A Case Study," *Science and Public Policy*, Vol. 45, No. 1, 2018, pp. 103–116.

Schmid, Jon, and Fei-Ling Wang, "Beyond National Innovation Systems: Incentives and China's Innovation Performance," *Journal of Contemporary China*, Vol. 26, No. 104, 2017, pp. 280–296.

Semiconductor Industry Association, "Strengthening the U.S. Semiconductor Industrial Base," webpage, undated. As of August 5, 2021:
https://www.semiconductors.org/strengthening-the-u-s-semiconductor-industrial-base/

Shor, Peter W., "Polynomial-Time Algorithms for Prime Factorization and Discrete Logarithms on a Quantum Computer," *SIAM Journal on Computing*, Vol. 26, No. 5, 1997, pp. 1484–1509.

Silberglitt, Richard, *New and Critical Materials: Identifying Potential Dual-Use Areas*, Santa Monica, Calif.: RAND Corporation, CT-513, 2019. As of August 5, 2021:
https://www.rand.org/pubs/testimonies/CT513.html

Smith, Thomas Bryan, Raffaele Vacca, Till Krenz, and Christopher McCarty, "Great Minds Think Alike, or Do They Often Differ? Research Topic Overlap and the Formation of Scientific Teams," *Journal of Informetrics*, Vol. 15, No. 1, 2021, p. 101104.

Smith-Goodson, Paul, "IonQ Releases a New 32-Qubit Trapped-Ion Quantum Computer with Massive Quantum Volume Claims," *Forbes*, October 7, 2020. As of August 5, 2021:
https://www.forbes.com/sites/moorinsights/2020/10/07/ionq-releases-a-new-32-qubit-trapped-ion-quantum-computer-with-massive-quantum-volume-claims/.

Sullivan, Mark, "How IonQ Is Planning to Bring a Quantum Computer to the Masses," *Fast Company*, October 1, 2021. As of August 5, 2021:
https://www.fastcompany.com/90682375/ionq-quantum-computing-going-public-spac

Tang, Ewin, "Quantum Principal Component Analysis Only Achieves an Exponential Speedup Because of Its State Preparation Assumptions," *Physical Review Letters*, Vol. 127, 2021, p. 060503.

U.S. Department of Commerce, Bureau of Industry and Security, Final Rule 86 FR 67317. As of December 13, 2021:
https://www.federalregister.gov/documents/2021/11/26/2021-25808/addition-of-entities-and
-revision-of-entries-on-the-entity-list-and-addition-of-entity-to-the

Valivarthi, Raju, Samantha I. Davis, Cristián Peña, Si Xie, Nikolai Lauk, Lautaro Narváez, Jason P. Allmaras, Andrew D. Beyer, Yewon Gim, Meraj Hussein, George Iskander, Hyunseong Linus Kim, Boris Korzh, Andrew Mueller, Mandy Rominsky, Matthew Shaw, Dawn Tang, Emma E. Wollman, Christoph Simon, Panagiotis Spentzouris, Daniel Oblak, Neil Sinclair, and Maria Spiropulu, "Teleportation Systems Toward a Quantum Internet," *PRX Quantum 1*, 020317, December 4, 2020.

Vermeer, Michael J. D., and Evan D. Peet, *Securing Communications in the Quantum Computing Age: Managing the Risks to Encryption*, Santa Monica, Calif.: RAND Corporation, RR-3102-RC, 2020. As of August 5, 2021:
https://www.rand.org/pubs/research_reports/RR3102.html

Wang, Brian, "Google on Track to Make Quantum Computer Faster Than Classical Computers Within 7 Months," *Next Big Future*, June 23, 2017. As of August 5, 2021:
https://www.nextbigfuture.com/2017/06/google-on-track-to-make-quantum-computer-faster
-than-classical-computers-within-7-months.html

Web of Science Group, "Web of Science Core Collection," *Clarivate.com*, webpage, undated. As of August 5, 2021:
https://clarivate.com/webofsciencegroup/solutions/web-of-science-core-collection/

Webb, James L., Luca Troise, Nikolaj W. Hansen, Jocelyn Achard, Ovidiu Brinza, Robert Staacke, Michael Kieschnick, Jan Meijer, Jean-François Perrier, Kirstine Berg-Sørensen, Alexander Huck, and Ulrik Lund Andersen, "Optimization of a Diamond Nitrogen Vacancy Centre Magnetometer for Sensing of Biological Signals," *Frontiers of Physics*, Vol. 8, 2020. As of August 5, 2021:
https://www.frontiersin.org/articles/10.3389/fphy.2020.522536/full

Wengerowsky, Sören, Siddarth Koduru Joshi, Fabian Steinlechner, Julien R. Zichi, Sergiy M. Dobrovolskiy, René van der Molen, Johannes W. N. Los, Val Zwiller, Marijn A. M. Versteegh, Alberto Mura, Davide Calonico, Massimo Inguscio, Hannes Hübel, Liu Bo, Thomas Scheidl, Anton Zeilinger, André Xuereb, and Rupert Ursin, "Entanglement Distribution over a 96-Km-Long Submarine Optical Fiber," *PNAS*, Vol. 116, No. 14, April 2, 2019, pp. 6684–6688.

White House, *Building Resilient Supply Chains, Revitalizing American Manufacturing, and Fostering Broad-Based Growth: 100-Day Reviews Under Executive Order 14017*, Washington, D.C.: White House, June 2021.

Wolchover, Natalie, "To Invent a Quantum Internet," *Quanta Magazine*, September 25, 2019.

Wright, K., K. M. Beck, S. Debnath, J. M. Amini, Y. Nam, N. Grzesiak, J.-S. Chen, N. C. Pisenti, M. Chmielewski, C. Collins, K. M. Hudek, J. Mizrahi, J. D. Wong-Campos, S. Allen, J. Apisdorf, P. Solomon, M. Williams, A. M. Ducore, A. Blinov, S. M. Kreikemeier, V. Chaplin, M. Keesan, C. Monroe, and J. Kim, "Benchmarking an 11-Qubit Quantum Computer," *Nature Communications*, Vol. 10, No. 1, November 29, 2019, p. 5464.

Wu, Xuejian, Zachary Pagel, Bola S. Malek, Timothy H. Nguyen, Fei Zi, Daniel S. Scheirer, and Holger Müller, "Gravity Surveys Using a Mobile Atom Interferometer," *Science Advances*, Vol. 5, No. 9, 2019, p. eaax0800.

Wu, Yulin, Wan-Su Bao, Sirui Cao, Fusheng Chen, Ming-Cheng Chen, Xiawei Chen, Tung-Hsun Chung, Hui Deng, Yajie Du, Daojin Fan, Ming Gong, Cheng Guo, Chu Guo, Shaojun Guo, Lianchen Han, Linyin Hong, He-Liang Huang, Yong-Heng Huo, Liping Li, Na Li, Shaowei Li, Yuan Li, Futian Liang, Chun Lin, Jin Lin, Haoran Qian, Dan Qiao, Hao Rong, Hong Su, Lihua Sun, Liangyuan Wang, Shiyu Wang, Dachao Wu, Yu Xu, Kai Yan, Weifeng Yang, Yang Yang, Yangsen Ye, Jianghan Yin, Chong Ying, Jiale Yu, Chen Zha, Cha Zhang, Haibin Zhang, Kaili Zhang, Yiming Zhang, Han Zhao, Youwei Zhao, Liang Zhou, Qingling Zhu, Chao-Yang Lu, Cheng-Zhi Peng, Xiaobo Zhu, and Jian-Wei Pan, "Strong Quantum Computational Advantage Using a Superconducting Quantum Processor," *ArXiv.org, Quantum Physics*, June 28, 2021. As of August 5, 2021:
http://arxiv.org/abs/2106.14734

Wuchty, Stefan, Benjamin F. Jones, and Brian Uzzi, "The Increasing Dominance of Teams in Production of Knowledge," *Science*, Vol. 316, No. 5827, 2007, pp. 1036–1039.

Yin, Juan, Yu-Huai Li, Sheng-Kai Liao, Meng Yang, Yuan Cao, Liang Zhang, Ji-Gang Ren, Wen-Qi Cai, Wei-Yue Liu, Shuang-Lin Li, Rong Shu, Yong-Mei Huang, Lei Deng, Li, Qiang Zhang, Nai-Le Liu, Yu-Ao Chen, Chao-Yang Lu, Xiang-Bin Wang, Feihu Xu, Jian-Yu Wang, Cheng-Zhi Peng, Artur K. Ekert, and Jian-Wei Pan, "Entanglement-Based Secure Quantum Cryptography over 1,120 Kilometres," *Nature*, Vol. 582, 2020, pp. 501–505.

Yin, Juan, Yuan Cao, Yu-Huai Li, Sheng Kai Liao, Liang Zhang, Ji-Gang Ren, Wen-Qi Cai, Wei-Yue Liu, Bo Li, Hui Dai, Guang-Bing Li, Qi-Ming Lu, Yun-Hong Gong, Yu Xu, Shuang-Lin Li, Feng-Zhi Li, Ya-Yun Yin, Zi-Qing Jiang, Ming Li, Jian-Jun Jia, Ge Ren, Dong He, Yi-Lin Zhou, Xiao-Xiang Zhang, Na Wang, Xiang Chang, Zhen-Cai Zhu, Nai-Le Liu, Yu-Ao Chen, Chao-Yang Lu, Rong Shu, Cheng-Zhi Peng, Jian-Yu Wang, and Jian-Wei Pan, "Satellite-Based Entanglement Distribution over 1200 Kilometers," *Science*, Vol. 356, No. 6343, June 16, 2017, pp. 1140–1144.

Yu, Yong, Fei Ma, Xi-Yu Luo, Bo Jing, Peng-Fei Sun, Ren-Zhou Fang, Chao-Wei Yang, Hui Liu, Ming-Yang Zheng, Xiu-Ping Xie, Wei-Jun Zhang, Li-Xing You, Zhen Wang, Teng-Yun Chen, Qiang Zhang, Xiao-Hui Bao, and Jian-Wei Pan, "Entanglement of Two Quantum Memories via Fibers over Dozens of Kilometres," *Nature*, Vol. 578, 2020, pp. 240–245.

Zhang, Eric J., Srikanth Srinivasan, Neereja Sundaresan, Daniela F. Bogorin, Yves Martin, Jared B. Hertzberg, John Timmerwilke, Emily J. Pritchett, Jeng-Bang Yau, Cindy Wang, William Landers, Eric P. Lewandowski, Adinath Narasgond, Sami Rosenblatt, George A. Keefe, Isaac Lauer, Mary Beth Rothwell, Douglas T. McClure, Oliver E. Dial, Jason S. Orcutt, Markus Brink, and Jerry M. Chow, "High-Fidelity Superconducting Quantum Processors via Laser-Annealing of Transmon Qubits," *ArXiv.org, Quantum Physics*, December 15, 2020. As of August 5, 2021:
http://arxiv.org/abs/2012.08475

Zhang, Qiang, Feihu Xu, Li, Nai-Le Liu, and Jian-Wei Pan, "Quantum Information Research in China," *Quantum Science and Technology*, Vol. 4, No. 4, 2019, p. 040503.

Zhang, Zheshen, Sara Mouradian, Franco N. C. Wong, and Jeffrey H. Shapiro, "Entanglement-Enhanced Sensing in a Lossy and Noisy Environment," *Physical Review Letters*, Vol. 114, 2015, p. 110506.

Zhijia, Lin, "Quantum Technology Commercialization Finds a Path, Capital Builds Momentum, but Technology Landing Pad Difficult," *Titanium Media APP*, in Chinese, March 16, 2021. As of August 5, 2021:
https://baijiahao.baidu.com/s?id=1694392230221672874&wfr=spider&for=pc

Zhong, Han-Sen, Yu-Hao Deng, Jian Qin, Hui Wang, Ming-Cheng Chen, Li-Chao Peng, Yi-Han Luo, Dian Wu, Si-Qiu Gong, Hao Su, Yi Hu, Peng Hu, Xiao-Yan Yang, Wei-Jun Zhang, Hao Li, Yuxuan Li, Xiao Jiang, Lin Gan, Guangwen Yang, Lixing You, Zhen Wang, Li Li, Nai-Le Liu, Jelmer Renema, Chao-Yang Lu, and Jian-Wei Pan, "Phase-Programmable Gaussian Boson Sampling Using Stimulated Squeezed Light," *ArXiv.org, Quantum Physics*, July 5, 2021. As of August 5, 2021:
http://arxiv.org/abs/2106.15534

Zhong, Han-Sen, Hui Wang, Yu-Hao Deng, Ming-Cheng Chen, Li-Chao Peng, Yi-Han Lui, Jian Qin, Dian Wu, Xing Ding, Yi Hu, Peng Hu, Xiao-Yan Yang, Wei-Jun Zhang, Hao Li, Yuxuan Li, Xiao Jiang, Lin Gan, Guangwen Yang, Lixing You, Zhen Wang, Li Li, Nai-Le Liu, Chao-Yang Lu, and Jian-Wei Pan, "Quantum Computational Advantage Using Photons," *Science*, Vol. 370, No. 6523, December 18, 2020, pp. 1460–1463.

Zhu, Qingling, Sirui Cao, Fusheng Chen, Ming-Cheng Chen, Xiawei Chen, Tung-Hsun Chung, Hui Deng, Yajie Du, Daojin Fan, Ming Gong, Cheng Guo, Chu Guo, Shaojun Guo, Lianchen Han, Linyin Hong, He-Liang Huang, Yong-Heng Huo, Liping Li, Na Li, Shaowei Li, Yuan Li, Futian Liang, Chun Lin, Jin Lin, Haoran Qian, Dan Qiao, Hao Rong, Hong Su, Lihua Sun, Liangyuan Wang, Shiyu Wang, Dachao Wu, Yulin Wu, Yu Xu, Kai Yan, Weifeng Yang, Yang Yang, Yangsen Ye, Jianghan Yin, Chong Ying, Jiale Yu, Chen Zha, Cha Zhang, Haibin Zhang, Kaili Zhang, Yiming Zhang, Han Zhao, Youwei Zhao, Liang Zhou, Chao-Yang Lu, Cheng-Zhi Peng, Xiaobo Zhu, and Jian-Wei Pan, "Quantum Computational Advantage via 60-Qubit 24-Cycle Random Circuit Sampling," *ArXiv.org, Quantum Physics*, September 9, 2021. As of August 5, 2021:
https://arxiv.org/abs/2109.03494